ENDPAPERS Burchell's zebra at a waterhole in Namibia.

OPPOSITE Jewel fish guarding its recently hatched and vulnerable fry.

OVERPAGE Greater flamingos in a bowing display near their breeding island, Lake Elmenteita, Kenya.

CONTENTS PAGE A procession of young antelopes, mostly immature females, across the savannah in East Africa.

The Family of Animals

The Family of Animals

Maurice Burton D. Sc.

ARCO PUBLISHING COMPANY, INC.
New York

Published by Arco Publishing Company, Inc.
219 Park Avenue South, New York, N.Y. 10003

Printed in Great Britain

Library of Congress Cataloging in Publication Data

Burton, Maurice, 1898–
The family of animals.

Includes index.
1. Zoology. I. Title.
QL50.B967 591 78–17855
ISBN 0–668–04662–7

Contents

Introduction

At a conservative estimate one million species of animals, of all shapes and sizes, live on the earth. The smallest are too tiny to be seen with the naked eye. The largest, the blue whale, the largest animal ever to have lived, is alive today, and could grow to a length of 100 ft (33 m) – if allowed to do so by the whaling industry.

There are animals living on land, in the sea, in lakes and rivers, as well as in the depths of the oceans and high on mountain tops. Animal species are to be found in cold climates and in hot, in swamps and in parched deserts. Some come out only at night, others by day. Every different environment whether it is grassland or forest, snow-covered waste or windswept rock, has its animal occupants.

Some animals are permanently fixed to a solid support, once their larval life is past. Others traverse the globe in their migrations. Some crawl, others fly; most have legs numbering two to a hundred. Some animals feed on plants, others prey on other animals. Yet others scavenge, helping to keep the world clean.

There are animals so simple in form that they are like flattened plates, while at the other end of the scale there are those so ornamented that they suggest the works of a celestial sculptor. There are animals that are drab and some coloured beyond the skill of the most talented artist.

Yet this vast assemblage of unlike creatures is truly one large family, united in the way that they live and in the things that they do. If this were not so, all would be utter confusion. As it is, they are built from the same basic substance, they live according to known rules, and these apply both to the plankton organisms with life-spans measured in a few hours and to those like the giant tortoise which lives for two hundred years.

Faced with this vast profusion, we shall proceed to sort out the pattern of these activities from the birth of the individual to the preparations for the next generation, to tell the story of the Family of Animals.

OPPOSITE The male frigate-bird with the bright red, balloon-like pouch he wears when courting. Frigate-birds spend most of their time in the air, over the sea and over land, rarely touching down on either. On land they can only shuffle clumsily or climb to a high place to take off. At sea they either scare other birds into disgorging their load of food or pick animals out of the water deftly with their bill. If, accidently, they land in the water they quickly become waterlogged.

1

The Miracle of Birth

When an Ancient Roman spoke of his *familia* he was referring to his household, and especially to his servants. During the ensuing two thousand years the word 'family' has come to mean many things. One meaning, for example, is everything descended from a common ancestor, and it was with this meaning in mind that the title of this book was chosen for we have reason to believe that all animals now living, and all that have lived, came from a common ancestor.

The word 'family', as we normally use it today meaning mother, father and children, did not come into use until the sixteenth century. It is fitting that this unit of parents and their offspring should be taken as our starting point.

The vast majority of animals start life as an ovum, or egg-cell, which is a tiny part of the mother's body. But reproduction, meaning the production of a new life, usually cannot proceed until another cell known as the sperm, from the father's body, has entered the ovum and fertilized it. The two cells, the sperm and the ovum, are brought together at mating. The moment they are actually joined is known as the moment of conception. It is the miraculous instant when a unit of protoplasm, which is the basis of life in animals and plants and is often microscopically small, starts on its own path of development. In due course it may become an individual animal as large as a six-ton elephant or a hundred-ton blue whale, or merely a tiny animal no bigger than a pin-head but able to carry on all the functions of life including reproducing itself.

Before going further some explanation may be helpful. 'Ovum' is simply the Latin for an egg, which is apt to cause confusion since what we normally think of as an egg is enclosed in a hard shell.

An ovum begins as an ordinary body cell which grows larger than the other cells; it is usually spherical and as it grows it becomes plentifully supplied with food material, known as yolk. In some ova (eggs) the amount of yolk is small. In others, as in birds' eggs, it is very large; and what is normally called the yolk of an egg is in fact the ovum distended by an accumulation of golden-yellow food material.

The size of the ovum, as a result of the varying amount of yolk, ranges from microscopic to several inches across. Probably the largest ovum is that of the ostrich, although the ova of the large pythons are only just smaller. The ova of the large sharks are possibly even larger. Most ova are 1/250 in (0·1 mm) in diameter and one of the smallest is the human ovum, which is about 1/2700 in (0·009 mm) in diameter.

It is the urge to reproduce that has carried on the stream of life from the beginning, thousands of millions of years ago, until the present and will carry it on into the future. It is the miracle of birth; but birth can take as many forms as there are different species of animals, past, present and future. For although all animals have so much in common, no two species live and behave completely alike and the manner in which they enter the world differs from species to species if only in small detail.

OPPOSITE Red deer hind with her calf at the moment of completion of birth. The first impulse of a mammalian mother is to inspect her newborn and start to lick it. In doing so she frees it of the remaining birth membranes, clears its nostrils so that it is able to breath properly and, continuing the action, licks her baby dry.

ASEXUAL REPRODUCTION

There must have been a time when there were no separate sexes, no male and female, and reproduction was wholly non-sexual. There were no

ova, and no sperms. This would, however, have been in the beginning, a very long time ago, but it still occurs in some animals and is called asexual reproduction to distinguish it from the better-known sexual reproduction in which two partners participate. There are also other forms of non-sexual reproduction, including budding, fission and fragmentation. In these no ova are formed. The animal produces offspring by giving up even larger parts of itself.

Virgin birth

The stick insect, sometimes called the walking-stick insect, has an incredibly long and slender body which is brownish in colour. It also has three pairs of long, slender legs. It feeds on leaves and lives on the shrubs bearing these leaves. When disturbed the stick insect folds its legs along its body, looking exactly like a twig or small stick, and seems to vanish.

Many stick insects reproduce by parthenogenesis, popularly called virgin birth. That is, the eggs are fertile although no male is present. In stick insects kept in the laboratory only one male was found for every four thousand females and in two New Zealand species no males have ever been found. The eggs have hard shells and look like seeds; they often look like seeds of the plant the insects eat. The female stick insect ejects several eggs a day which fall to the ground where they hatch, weeks or months later. Where the insects are numerous the falling eggs can patter through the leaves with the sound of rain.

Virgin birth is common in many insects and the smaller crustaceans. In one instance a culture of small crustaceans was kept under observation in the laboratory for several decades. At no time was a male seen in the numerous generations.

In normal fertilization the nucleus of the ovum carries the characters – that is, the qualities and features – of the mother, that of the sperm carries the characters of the father. The entry of the sperm into the ovum is followed by its nucleus fusing with the nucleus of the ovum. So the fertilized ovum contains the characters of the two parents intermingled. The ovum then divides into two; these two halves then divide into four, the four divide into eight and so on until a mass of daughter cells is

Shortly before egg-laying a pair of freshwater jewel fishes clean the surface of a rock or large stone, removing all small plant growths. On this clean patch the female lays her eggs and the male fertilizes them by shedding his milt over them. Both parents keep the eggs clean, taking it in turns to fan them with their pectoral fins. This keeps the water circulating so that the eggs are adequately supplied with oxygen. The parents protect the fry also until about a week after they have become free-swimming.

formed. This becomes the embryo from which the future animal is born. Consequently, every offspring has a mixture of the characters of the maternal and the paternal parents. In virgin birth, however, the offspring carries only the characters of the mother.

Virgin birth is most common among invertebrates, the lower animals that lack a backbone. It may also occur, but rarely, in vertebrates, animals with a backbone, but there was the sensational discovery in 1966 that rock lizards of the Caucasus and whiptail lizards in North America reproduce by virgin birth. It was sensational because nobody had ever supposed that virgin birth could occur in the higher animals. More surprising still, turkeys on turkey farms have been known to lay fertile eggs without mating and abortive parthenogenesis, that is virgin birth which does not extend beyond the development of an embryo, has been detected in some mammals, the highest and most specialized class of the animal kingdom.

How virgin birth comes about is still something of a mystery. In the 1920s it was announced that a scientist had made sea-urchin eggs divide and grow into larvae by merely pricking them with a fine needle. In further experimentation it has been found that the eggs of silkworms, sea-urchins, marine worms, starfishes and frogs can be made to develop into larvae by the same method. The same result can be obtained by subjecting the ova to abrupt changes of temperature, to shaking or to immersing them in solutions of acids or alkalis or in solutions of salt. The press coverage at the time created a mild sensation and, remembering that sea-water is a solution of salt, some ladies became apprehensive about bathing in the sea!

We can now be certain that these treatments give a stimulus that causes the ovum to start dividing, so presumably the sperm not only contributes the father's characters but merely by entering the ovum, the equivalent of pricking with a needle, causes it to divide. We also know that sometimes a sperm will enter an ovum and fail to connect with the nucleus of the ovum, yet the ovum will divide as usual. Although in these experiments the larvae failed as a rule to reach maturity, we can presume they would, had they survived, all have given rise to females only.

Females reproducing by virgin birth lay more eggs than those that have mated. Rock lizards illustrate this: in some parts of the Caucasus there are as many males as females but only in certain parts of their range are the rock lizards represented by females only, and these lay a larger number of eggs by virgin birth than those that have mated.

Plantlice also illustrate this well; these are the insects known variously as greenfly and blackfly and collectively as aphides. They are pests found on rose bushes and vegetable crops, and live by sucking the sap from them. All through the summer plantlice are all females, some wingless, others with wings, that can spread from place to place. Up to fifty babies are born by virgin birth to each female.

With a magnifying glass one can watch each baby being born. It emerges from the tail end of the mother as a small green lump that grows bigger and longer. Then the legs grow out and slowly wave about and in ten to twenty minutes the baby drops to the surface of the leaf, pushes its beak-like proboscis into the skin of the leaf and starts to suck sap. In a week or so it will be giving birth to its own babies. In some plantlice the babies are born with their own babies inside them almost ready to be born in turn.

With regard to their rate of reproduction and their growth it has been estimated that 'if all the progeny of a single female aphid survived they would, in the course of a year, equal the weight of 500 million stout men'. As it is, plantlice have so many enemies that most die early. The larvae of ladybirds, lacewing flies and hoverflies, as well as small birds, devour them sufficiently to keep their numbers within bounds.

The armadillo conceives its young by a combination of sexual and asexual reproduction. Armadillos as a family are strictly South American except for the nine-banded armadillo, shown here, which has surprisingly extended its range northwards to the southern United States.

Alternation of method

Plantlice regularly alternate virgin birth with sexual reproduction. In the autumn the females lay eggs instead of giving birth to live young. From some eggs hatch females, from others come males. Mating takes place, fertilized eggs are laid and these hatch the following spring, giving the females that start the summer-long reproduction of live young by virgin birth all over again.

There is another way in which asexual reproduction can alternate with sexual reproduction. In some species of wasps that are parasites on the larvae of other insects the fertilized ovum begins to develop into an embryo soon after it is laid. This then begins to bud off groups of cells each of which develops into an embryo. In one species of wasp, as a result of this process, as many as 1500 larvae may be formed from a single fertilized ovum.

A similar process is seen in other animals, for example the armadillos of the warmer parts of America. As the fertilized ovum divides, the parts separate and each develops into an embryo, and from a single ovum there may come as many as twelve embryos. Not all survive, however. The usual litter is two, three or four, although sometimes only one embryo may survive. Where there is more than one, all are of the same sex like identical twins.

Identical twins represent a similar phenomenon. The fertilized ovum divides into two. The two halves separate and each develops into an embryo to give two babies that look alike for the rest of their lives.

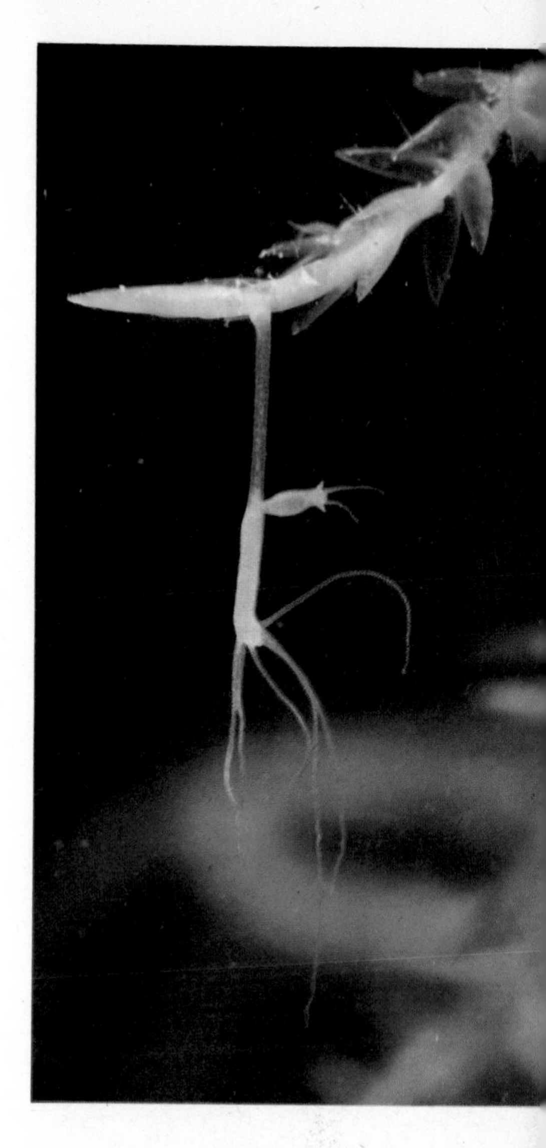

A brown hydra, related to sea-anemones but living in fresh water, with a bud growing from its side. In due course the bud will become detached from the parent, float away and fasten itself to a water plant. Sometimes a hydra may be seen with three or four buds in differing stages of development.

Budding

When a female gives rise to an ovum she is donating part of herself to the next generation. There are other ways of achieving the same end and of doing so without the help of a second parent, apart from virgin birth. This is by the various kinds of budding. In this, instead of there being the one cell we call the ovum, a group of cells is involved.

A typical budding is seen in the well-known hydra with which successive generations of biology students have become so familiar. Hydra is a small freshwater animal, green or brown, a relative of the sea-anemone and, like the anemone, it has a columnar body with a mouth at the free end surrounded by a circle of tentacles.

When there is an abundance of food material in the water, a bump appears on the side of the hydra and grows out like a finger. Tentacles appear near the free end, then a mouth, inside the circle of tentacles. At first this extraordinary baby draws nourishment from its parent's body. Then the tissues at its base contract, the new hydra is pinched off, drops away and fastens itself to another part of the aquatic plant on which its parent is living, and begins a new life independent of the parent.

Fission

Sea-anemones show the greatest possible range of asexual reproduction although not all the methods occur in any one species. Some species split lengthwise to produce two, sometimes more, individuals. In contrast to this longitudinal fission, as this process is called, other species use transverse fission. In this, a second ring of tentacles grows out half-way or more down the body, which then splits in two just above the line of tentacles. The top and bottom parts separate. The split surfaces heal, the missing parts are regenerated, and there are now two anemones. Sea-anemones also use a further method of reproduction.

Fragmentation

This further method of asexual reproduction is known as laceration or fragmentation. In some sea-anemones young ones arise at the base of the old one and become constricted off. They then move away or else the parent anemone wanders away from them, moving slug-like but more slowly than a slug, on its base. Some anemones, although usually stationary, at times start to roam and as they do so they leave pieces of the base behind, each piece re-forming to produce a baby anenome.

These two starfishes have grown from arms amputated from another starfish. Provided there is a piece of the central disc attached to the arm complete regeneration will take place, but pieces broken off an arm without any central part of the body, will not regenerate.

Other animals can reproduce similarly, especially sponges, which are lower in the animal scale than sea-anemones. A living sponge can be cut into small pieces each capable of growing into a new sponge. Attempts have been made, particularly off the Bahamas in the early years of this century, to grow bath sponges for market in this way. The sponges were taken from the sea, each cut into several pieces and fixed to small concrete discs. The sponge fragments were then returned to the sea floor, laid out like rows of cabbages. Unfortunately, however, the experiment failed and the scheme was ended.

Pieces of sponge broken off by strong wave action will regenerate to form new sponges but there is at least one kind that naturally sheds pieces of itself. This is the purse sponge, which is like a flattened goblet, white to yellow, averaging 2 in (50 mm) in height, that is found in great abundance on rocky shores in many parts of the world.

At certain times of the year, purse sponges develop grooves across the body. They split along the grooves, the outer segment floating away or dropping to the bottom, where it grows a new stalk. Sometimes a line of holes appear across the body, like the perforations of postage stamps. Eventually the outer part tears away and a new sponge is born.

Certain flatworms, only slightly higher in the animal scale than sponges and sea-anemones, can be cut into a dozen pieces. Each will re-form the missing parts to produce a dozen new flatworms.

That this fission is not accidental can be best indicated by the way in which a sea-anemone will play tug-of-war with itself, one half moving in one direction, the other moving in the opposite direction. Eventually the two are torn apart, heal their wounds and become two whole and complete sea-anemones. A starfish torn apart accidentally will grow into two, by the renewal of lost parts. Certain brittlestars, close relatives of starfishes, emulate the tug-of-war of anemones, as a normal method of increasing their numbers.

Some years ago a reef in the Caribbean was discovered to be populated by male brittlestars only. Presumably a male brittlestar had wandered on to the reef and split itself in two, and these were all its bachelor descendants!

Gemmules

Perhaps the most remarkable of all the asexual methods of reproduction

is that which involves the formation of gemmules. There are a number of plants that form special buds that drop away to form new plants. These are known as gemmae, which is Latin for buds. Gemmules, or little buds, are found especially in freshwater sponges and the so-called moss-animals.

A gemmule is formed as follows. Scores of ordinary body cells leave the tissues of which they were hitherto a part. They may come from every part of the body and, under the microscope, can be seen streaming in one direction towards an assembly point. There they gather together into a rounded mass. Other cells in the tissues immediately surrounding them lay down a tough, almost horny coat enclosing them.

During the winter the sponge or moss-animal dies. The cells within the tough capsule can stand up to freezing temperatures or drying out. The following spring the group of cells flows out of the capsule and grows into a new animal.

EGG-LAYING AND LIVE BIRTH

Although sea-anemones use such different methods of asexual reproduction they also reproduce sexually, and use two methods. Most sea-anemones are male or female, although in some species all are hermaphrodite, combining male and female in one individual. In some species, females shed their ova into the sea and the males shed their sperms, and in due course the ova are fertilized. In other species, the sperms shed into the water find their way into the female through the mouth and there fertilize the ova. One of two things may happen to fertilized ova. They may be extruded to complete their development in the sea, and this is known as oviparity or egg-laying. In the other method the fertilized ova develop inside the maternal body and are miniatures of the parent when they make their exit. This second method is known as viviparity, or live birth.

Birth of a salamander. *Left*, the parents with a cluster of eggs, each egg contained within its spherical envelope of jelly which not only buffers the developing embryo from injury but insulates it from the cold. The eggs grow into embryos within the protective jelly nourished from a store of yolk contained within the original ovum (*centre*). These formless embryos will eventually come to resemble an adult salamander while still within the jelly coat. In the early stages of development the embryo lies quiescent, but as the limbs and tail reach their mature form muscular movement is initiated from the base of the head and the embryo begins to wriggle, finally bursting the surrounding envelope. The young salamander emerges complete with gills on its neck (*right*). In due course the young salamander grows a pair of lungs after which the gills shrivel and finally disappear.

OPPOSITE A female European adder with her young, which are born live, but hatched from eggs. Some animals bear live young. Others lay eggs and the young are hatched outside the mother's body. The adder, in common with some other reptiles, does both. The eggs develop inside the mother's body but hatch as they are being laid so that the babies leave her body live.

The most common method of reproduction, taking the animal kingdom as a whole, is the sexual method. Asexual reproduction can be almost as common in some of the lower levels but it becomes less frequent as we go up the scale to the more highly organized animals. Apart from the few exceptions already noted, asexual reproduction does not occur among the higher animals.

Egg-laying and live birth are found throughout the animal kingdom, and of these two, egg-laying is the more frequent. It reaches its highest expression in birds, every species of which lays eggs. Live birth, on the other hand, is a characteristic of all but two of the most highly organized of animals, the mammals, or furred animals. The exceptions are the platypus or duckbill and the echidna or spiny anteater, both of Australasia. They are the egg-laying mammals.

Occasionally there is a process intermediate between egg-laying and live birth. In this the ova are retained in the maternal body until they hatch. Although eggs are formed it looks like live birth and, accordingly, this process is called ovoviviparity. In certain snakes and in the European slowworm, a legless lizard, the eggs hatch at the last moment and the young enter the world without the membrane that represents the shell. Not infrequently there is a slight delay in the hatching and the baby reaches the exterior still inside its membrane. Hatching takes place soon after the egg has left the mother's body.

Egg-laying is seen at its simplest in aquatic animals. The females merely shed their ova into the surrounding water, the males shed their sperms, and so long as the two actions are simultaneous the sperms will find their way to the ova and fertilize them. The sperms are very small, very active and are usually tadpole shaped, with an oval-shaped head and a long, slender tail. A sperm swims by waggling its tail and is attracted by a chemical given out by the ova. External fertilization, as this is called, is used by most of the twenty thousand species of fishes, as well as by frogs, toads and newts.

Nevertheless, even in animals living permanently in water, there are many instances of internal fertilization, the sperms swimming towards the female and entering her body through an appropriate opening. For land-living animals other than amphibians internal fertilization is essential. The male then needs some other method of conveying his

BELOW A green lizard newly hatched from the shell. The other three eggs are yet to hatch. The eggs of reptiles are typically enclosed in a parchment-like shell which the baby slits with an egg-tooth growing on the tip of its snout in order to escape from the shell.

BELOW RIGHT A male midwife toad carrying a string of eggs entwined round his hindlegs. He visits water from time to time to immerse his legs in order to prevent the eggs drying up, and when the eggs are due to hatch he again visits water but this time for the emerging tadpoles to enter it.

A nestful of ostrich chicks with some eggs still unhatched. The male ostrich is polygamous and up to five hens lay their eggs in the nest. The male does most of the incubating but is relieved by one or other of the hens during at least part of the daylight hours.

sperms to the ova. This is done by use of an organ capable of insertion, which may take a variety of forms.

Eggs laid in water require little protection compared with those laid on land. They cannot dry out, and the water acts as a buffer to protect them from being readily smashed. Eggs laid on land need protection from the sun and the drying effects of the wind. They also need protection against being smashed accidentally, so usually they have hard, resistant shells.

The make-up of an egg

The egg-cell or ovum is basically like all other living cells in its structure. It is made up of a nucleus surrounded by cytoplasm – a collection of certain complex organic compounds within the cell – the whole enclosed by a limiting membrane. The nucleus carries the chromosomes bearing the genes that govern what sort of animal the fertilized ovum will become and how it will behave. The cytoplasm usually contains yolk, although the amount varies with the species, as we have seen, and in some instances, for example in many flatworms, the yolk is formed but is not enclosed in the egg-cell.

All egg-cells are enclosed in a membrane. In insects there is another outer layer which is added as the egg is laid. This hardens and its surface is sculptured often with intricate and beautiful designs. In sharks, skates and rays, the egg-cell, as it is being laid, becomes enclosed in a leathery shell. This, when found cast up on the beach, is known as a mermaid's purse. Reptiles' eggs are covered with a parchment-like shell.

As birds' ova, heavily charged with yolk, are passing down the oviduct, the tube that leads to the outside, they become coated by special glands in the wall of the oviduct with a layer of albumen, the white of egg. Outside this two membranes are laid down by further glands. These are firmly fastened together except at the broad end where they separate to form an air-space. Outside these a coating of lime is laid down as the egg slips farther down the oviduct. This is the shell proper which is made up of several layers, each with its own microscopic pattern of tubes and channels so that gases can pass through. The pigments giving the colours to the egg are the last to be added before the egg is laid.

The embryo grows

An egg laid on land must have a complicated structure since it has a lot of work to do. Apart from keeping the egg warm and trying to protect it from enemies, the parents can do nothing to help the growing chick inside the egg. It must be largely dependent on its built-in survival kit, as represented by the yolk, white, membranes and shell. It needs to breathe and it can take in oxygen through the several layers that surround it, and the spent air, in the form of carbon dioxide, a waste substance, is got rid of by the reverse route. It needs to feed and yolk is provided but there are waste substances from food. In some eggs these are converted to uric acid, which is insoluble in water and so it does not get into the bloodstream of the embryo. In reptiles and birds, waste in the form of urea, which is soluble in water, accumulates in a sort of bladder and is left behind in the shell on hatching.

The growing embryo must be buffered against sudden jolts and it is protected by the cushion of albumen that surrounds it. In addition there are the membranes and shell, and all not only give protection against mechanical damage but prevent the future infant from drying up.

In mammals, other than platypus and echidna, the mother's body takes the place of the shell. The embryo, called in mammals the foetus when it begins to take recognizable shape, is fully protected from the outside world. It receives food and oxygen from the mother's blood, and its waste products are carried away also in her blood and excreted through her organs.

When a chick hatches

Hatching is something that does not suddenly happen. There are preparations that precede the breaking of the shell. In birds' eggs, the most complex of all eggs, these preparations may last from half an hour, in the eggs of small birds, to several hours for the larger species. Their duration is, however, fixed for each species.

Two of the most vital processes are that the chick should break the shell, to make its exit, and that it must become air-breathing. Up to this point it has obtained its oxygen from air passing through the porous shell

Chick of the domestic hen about to hatch. It has broken the eggshell using the egg-tooth which can be seen as a tiny bumb on its beak. The breaking of the shell is not accomplished in one movement. The chick moves its head in a nodding manner, at the same time moving it sideways, and so taps a series of holes which weaken the shell. Then by stretching movements of the body and legs the shell is finally ruptured and the chick emerges.

and entering the blood-vessels. Throughout incubation there has been a slight but steady loss of water by evaporation. This causes an air chamber to form between the two membranes lining the shell, at its blunt end. The chick pushes its bill into this to take its first breath. This is also the moment when it makes its first chirping and it may occur several days before the shell begins to crack. The chick is still receiving oxygen into its blood as before.

Meanwhile, the chick swallows any liquid left in the egg, including the remains of the white. What is left of the yolk, which may be anything up to a quarter of the weight of the chick and is contained in a yolk-sac, is drawn by muscular action through the navel and into the chick's body.

Exit from the shell is achieved by three things. Firstly, the shell has grown progressively thinner and weaker during the course of incubation. Secondly, early in development a small horny 'thorn', known as the egg-tooth, grows on the upper part of the beak near its tip; this is used as a kind of pick or awl to make holes in the shell. Thirdly, after the first openings are made with it, the movements of the chick's legs and the stretching movements of its head and neck help to burst the shell in a neat and orderly fashion. The egg-tooth drops off soon after the chick has hatched.

Reptiles also have an egg-tooth which is like a real tooth but has a razor-sharp edge that cuts through the parchment-like shell. Geckos have two egg-teeth. Other animals hatching from eggs burst the shell by the vigorous movements of the body, the rupture taking place along a line of weakness in the shell.

The first chirpings of the chick alert the hen to what is happening. In crocodiles, the only reptiles that make a nest and assist the hatchlings to leave it, the first sounds cause the mother to start loosening the top of the covered-in nest.

PART REPTILE, PART MAMMAL

The great majority of animals are what is popularly called cold-blooded. More precisely, they should be called poikilothermic, a compound of two Greek words that means approximately that their temperature varies with that of their surroundings. A reptile, for example, exposed to blazing sunshine becomes temporarily warm-blooded. By contrast, birds and mammals are warm-blooded, strictly speaking, homoiothermic, meaning that their temperature remains constant whatever the temperature of the surrounding air.

It has been the custom to bracket birds and mammals as distinct from the rest of the animal kingdom because they are the only warm-blooded animals. Today, there is a tendency to view birds as more nearly related to reptiles than to mammals; indeed, some newly hatched baby birds, pelicans are a particular example, look very reptilian.

By a fortunate chance, there have been preserved from extinction two mammals that serve as a link with reptiles. They are the two egg-laying mammals already mentioned, the echidna and platypus.

When we speak of the family of animals, using the word in its broader sense, namely, of all animals being descended from a common ancestor, we imply that all have evolved from a single species far back in time. The idea is not acceptable to everyone, for there are those who prefer to believe in a separate creation for each species. It is true that the theory of a continuous evolution has many weak links in its chain of evidence. These are, however, counterbalanced by some surprisingly convincing strong links elsewhere.

The adult echidna and platypus supply one such link. They are mammals in that their body-covering is hair and that they suckle their young. Yet many of their bones, and other features of their anatomy, are distinctly reptilian in style. They are also reptilian in that they lay eggs with parchment-like shells. They are also similar to reptiles in the way the young develop within the shell and in having an egg-tooth for ripping open the shell.

Both are mammalian in that once the egg hatches the baby feeds on milk that oozes from patches of glands on the mother's belly. The platypus lays two small eggs and incubates them in an underground nest. The echidna lays one egg only and transfers it – no one knows how – to a pouch on her abdomen where it hatches ten days later. The baby remains in the pouch until its spines become an inconvenience to the mother.

POUCH-BEARERS

The use of a pouch for brooding the eggs or young is by no means unique to echidna. Throughout the animal kingdom are found isolated instances of species that use a brood-pouch. Most sea-urchins, for example, shed their ova into the sea, there to be fertilized. A few have a brood-pouch in which the eggs are held and the young remain, sheltered until old enough to look after themselves.

Out of the twenty thousand species of fishes only a few use a brood-pouch, notably the seahorses. At breeding time a pouch appears on the belly of the male. At the same time the female grows a long tube, or ovipositor, which she inserts into his pouch to lay her eggs. The male's

The platypus or duckbill of Australia, an egg-laying mammal, swimming in a river. In preparation for laying her eggs the female tunnels into a bank and prepares a nest. She curls her body around the two eggs to incubate them. When the young platypuses hatch they are fed with milk oozing from two glands on the mother's abdomen.

ABOVE A female wildebeest, a form of African antelope, with her newly-born calf that has just got up on to its feet within an hour of being born. All hoofed mammals are herbivores and therefore the prey of flesh-eaters. Most of them live by grazing or browsing in the open and it is essential that the baby should be able to run and keep up with the mother within a few hours of being born.

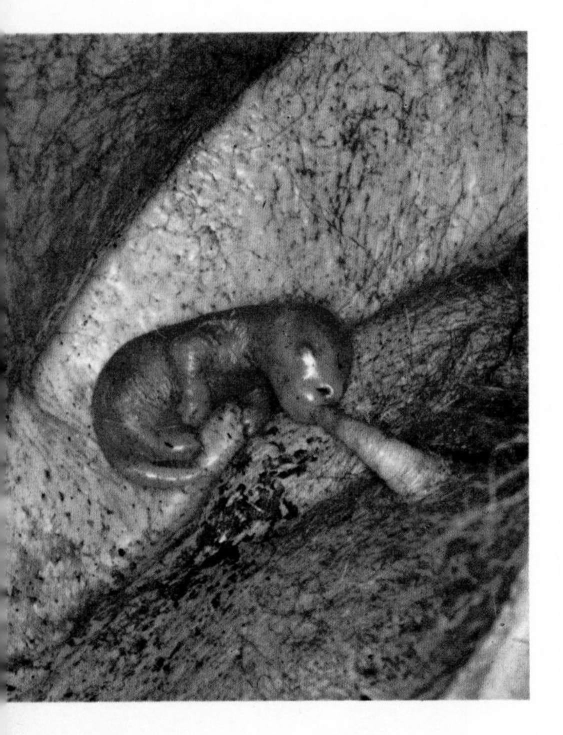

Baby kangaroo soon after birth lying in the mother's pouch and attached to a teat.

sperms are shed into the pouch. Mechanically, this is a reversal of the usual role of the sexes. Moreover, when the young have grown into miniature seahorses, it is the male that endures the exhausting ordeal of their birth. He expels each baby from the pouch with a convulsion comparable to a uterine spasm and when all the babies have been expelled (born) he lies limp for a while before recovering his strength and swimming away.

There is, however, one group of mammals that are consistently pouch-bearers: these are the marsupials, mainly found in Australasia today, with a few in America, although fossils show that they once extended throughout the world. Marsupials are familiar to all through the kangaroos, wallabies, koala, opossums and others.

In marsupials some of the features of anatomy are reptilian, although these are fewer than in the echidna and platypus. The female has a pouch on the abdomen. The ovum is retained in her body after mating but not for long. The baby is born while still a half-developed foetus, no larger than a haricot bean even when the mother is a 6-ft (1·8 m) tall kangaroo. Once born, this immature mite claws its way through the mother's fur to reach the shelter of the pouch and there seizes a teat in its mouth. It remains, firmly holding on to the teat, for most of its infancy, and continues to use the pouch as a shelter even when already on the way to becoming adult, in appearance at least.

THE TRUE MAMMALS

The marsupials are half-way between the egg-laying mammals and the true or placental mammals, the furred animals familiar to us in dogs and cats, farmstock and many others. They are called the placental mammals because the growing baby is contained within a fleshy bag known as the placenta, or after-birth, which is ejected by the mother soon after the baby is born.

In mammals the ovum is retained in the female's body after mating.

A common zebra has just foaled: the hindquarters of the baby are still enveloped in the birth membranes. Partly by its own efforts but mainly by the mother's licking it will soon be freed of the membranes and making its first efforts to get on to its feet.

After fertilization, it divides into the usual two cells, the two into four, the four into eight, and so forth. As the number of cells increases they first form a hollow sphere that becomes implanted in the wall of the uterus or womb. As further divisions of the cells occur, the future organs of the body, the gut, the nervous system, the blood-vessels, bones, and so on, gradually take shape forming the embryo. The embryo becomes a foetus as the developing baby takes on a form resembling the parents.

The placenta is highly important. It is rich in blood-vessels that are continuous with those of the mother. The blood flowing from mother to foetus brings in food and oxygen, and as it flows back into the mother's body it carries waste which is disposed of by her organs. During the gestation period, from conception to birth, mother and baby are one. This period may be only three weeks in a shrew, the smallest mammal, or nearly two years in an elephant.

Within the placenta, the developing baby is kept warm and buffered against injury. It is fed and cosseted, part of its mother yet preparing for life apart in the future, following a time-table laid down by the genes in its chromosomes, which have been derived in part from the father and in part from the mother.

The foetus is joined to the placenta by a tube called the umbilical cord which at birth must be snapped, its scar becoming the navel (umbilicus) that it retains throughout life. In due course gestation ends, the placenta ruptures releasing the fluids that have cushioned the foetus. Strong convulsions in the appropriate muscles of the mother send the foetus on its journey down the birth-canal. The baby emerges, head first, except in whales, to take its first breath. The umbilical cord snaps, or is severed by the mother's teeth. A new baby is born, a new individual that will one day give part of its own body to continue the succession of generations that make the animal kingdom one large family. The continuing convulsions of the mother, the tail-end of the labour pains, eject the placenta and birth is completed.

2
The Infancy of Animals

For the vast majority of the members of the animal kingdom, the laying of eggs or the birth of young sees the end of parental responsibility. In less than half of 1 per cent of the total number of species is there any form of parental care. In this vast majority, the offspring is left to its own devices, to take a chance on surviving the numerous hazards. As a result there is a high percentage of mortality, particularly from being eaten by other animals. Indeed, most of the animal kingdom provides reproductive material to feed a large part of the remaining species.

In any species a parent needs only two surviving offspring to ensure that the population remains stable. The American oyster is reputed to lay 500 million eggs a year, of which only one needs to survive because each of the edible oysters at least is known to be alternately male and female. Even among birds that lay relatively few eggs a year, around four to five on average and seldom more than twelve, 60–70% will have perished, either as eggs or as young birds within three to six months. Even in mammals, with rare exceptions, two out of every three babies born will not survive infancy.

Among the invertebrates, especially those wholly aquatic, production of large numbers of eggs is the rule. In marine species, the numbers of eggs laid per year are in tens or hundreds of thousands, sometimes in millions. The fate of the eggs, and the larvae hatching from them, can be illustrated by reference to krill, the prawn-like crustacean on which the great whales feed in the Antarctic.

Krill lay their eggs at the surface of the sea in late summer. Soon the eggs begin to sink to depths of as much as 6500 ft (2000 m). When the eggs hatch the larvae swim to the surface and take about two years to grow into adults. During this time they are carried around in shoals, some of them enormous, in ocean currents. Some of the shoals measure hundreds of yards across and contain hundreds of tons of krill. They form the food of many animals. The great whalebone whales feed almost exclusively on krill and it has been estimated that before their numbers were reduced by whaling, these whales ate 150 million tons of krill each year. Another estimate puts the amount of krill produced each year at 500 million tons. Krill is also eaten by fishes and squid, seabirds, especially penguins and albatrosses, and seals, especially the crabeater seal.

A similar picture could be drawn for almost all of the tens of thousands of marine invertebrates, as well as the marine fishes. The numbers of eggs laid in fresh waters tend to be much less, as do those laid on land. At certain times of the year the sea is crammed with eggs and larvae, all but a small percentage of which are destined to die by accident, adverse conditions of weather or, more particularly, by being eaten.

Yet although the great majority of animal parents do little to help or safeguard their offspring once the eggs are laid, there are a few exceptions even among the lower animals. Out of the million or so kinds of living insects only a handful even protect the eggs and fewer still protect the young. Earwigs are a notable exception. The female lays her eggs in a hollow in the ground and not only covers them with her body but turns them over and licks them to keep them clean. She also stays a few days with the larvae that hatch from them but does not feed them nor deliberately tend them.

OPPOSITE Three stages in the life of the clawed toad. The tadpole (below) has begun to grow its hindlegs. These are as yet weak and functionless and the tadpole still swims by wriggling its tail. As the hindlegs grow larger and stronger the forelegs make their appearance. The tail is then absorbed into the body until finally we have the complete change from a tadpole to a froglet (centre).

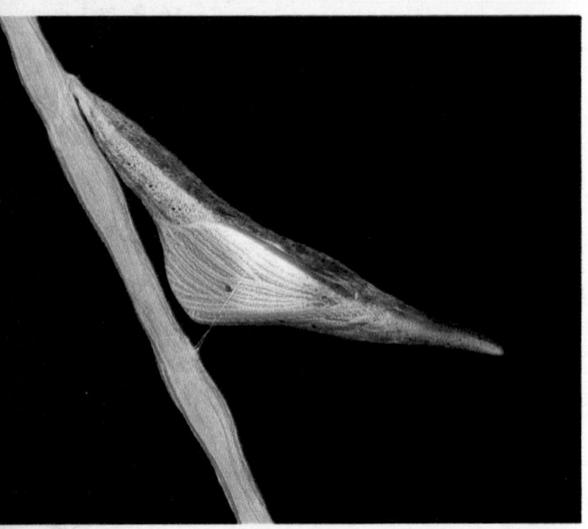

Life history of the orange tip butterfly.

TOP The larva feeding on pods of a cruciferous plant. When ready to pupate the larva fastens itself to the stem of the food plant by its rear end and spins a loop of silk from its body to sling it like a hammock. CENTRE The larva has now changed to a pupa and the form of the future butterfly can be seen through the pupal skin. In due course the butterfly leaves the pupal skin and hangs from it while its wings expand and dry. BELOW The adult butterfly ready to take off.

There are a few timber beetles, living in tunnels in growing trees, that not only feed their larvae with a fungus but clear away their droppings. Other beetles that live in decaying wood go further. Both male and female chew wood and feed it to the larvae. They also tend them throughout their larval life and help them build their cocoons when the larvae are ready to pupate.

There are spiders, known as wolf spiders, whose females lay their eggs in silken cocoons which they carry fastened to the tip of the abdomen. When the spiderlings hatch out they clamber on to the mother's back and are carried about during the vulnerable stage of their early lives. A few other spiders live in small communities in a web and these actually feed insects to their spiderlings.

Certain millipedes are dedicated mothers up to a point. The female constructs a small volcano-shaped chamber of mud, lays her eggs through the crater, then coils her body around the 'volcano' to protect the eggs. She continues this protection when the babies have hatched out.

Parental care in vertebrates

Most of the vertebrates do little more than the lower animals. There are a very few fishes that make a nest for their eggs or lay them in a rounded mass and then protect them either by curling around their egg-mass or by driving intruders from the nest. Sticklebacks are famous for the way in which the father builds a nest of soft water-plants, then he guards the nest and aerates the eggs by fanning them with his fins, to drive currents of oxygenated water over them. He also guards the nest after the eggs are hatched, watching over the young fry to see that they do not leave the nest too soon. Should one stray he swims after it, takes it into his mouth and spits it back into the nest.

Some tropical freshwater fishes are mouth-brooders. That is, male and female take the eggs, as they are laid, into their mouths to brood them. When the young hatch they are allowed to swim free but the moment danger threatens the babies swim back, the parent opens its mouth and they swim in to shelter until the danger is past.

Amphibians, including frogs, toads, newts and salamanders, usually lay their eggs in water and then depart, leaving them to their fate. Indeed, the only interest most frogs have in their babies is to snap them up and swallow them if the opportunity presents itself. There are, however, tropical tree frogs that build a foam nest in which to lay their eggs, but they do no more for their offspring than this.

When Surinam toads pair the female pushes out her tubular oviduct, which is the canal through which the ova leave the ovary. The male, in clinging to her back as he sheds his sperms, presses the oviduct on to her back and, as each egg is laid and fertilized, it is pressed into her skin which forms a pocket around it. Then a lid of skin grows over the pocket. The eggs hatch and later, from each pocket, there comes a tiny toadlet without tail or gills and it swims away.

The females of the pouched frogs have a pouch that extends over the whole of the back. When mating, the male pushes each egg as it is laid into the pouch with his hindfeet. In due course tiny froglets can be seen leaving the pouch.

Mother of millions: the queen termite, who spends her whole life laying eggs, imprisoned in a central chamber in the termite nest, where she is fed and tended by the workers. Her enormously distended abdomen dwarfs her head and thorax, seen on the left.

These few examples represent the limit of parental care in amphibians. Reptiles do little better for apart from making nests into which they lay their eggs, the only kinds that show any degree of parental care are the crocodiles; although some snakes coil their bodies around their eggs, thus protecting and incubating them.

It has been known for a long time that crocodiles and alligators make a nest in which to lay their eggs. Some dig a pit and cover the eggs with sand; others cover them with rotting vegetable matter. Then the mother stations herself near the nest. She becomes irritable and is prone to attack anything approaching the nest. When the baby crocodiles hatch out they make a kind of cheeping noise. The mother hears this and loosens the covering of the nest. Should she be absent, the young dig their way out and it is then, in the absence of the mother, that they may fall victim to several enemies.

It was discovered a few years ago that the mother crocodile does still more to care for her young. As the first youngsters emerge she takes them in her mouth and carries them to water. This provides protection during what would otherwise be a perilous journey for the hatchlings. More surprising, it was seen that, once in water, she opened her mouth slightly and shook her head. This had the effect of washing the babies. The mother did this slowly and carefully and, having washed the first babies, she went back to the nest for more, the young crocodiles waiting for her to fetch them. Perhaps the most surprising observation was that sometimes the father as well would assist in transporting and washing his young offspring.

However, nothing so far mentioned compares with the parental care shown by birds and mammals, which feed, tend, protect and, in mammals, clean their offspring throughout their infancy. When one speaks of maternal devotion in animals one has in mind birds and, more especially, mammals.

Hen nightingale returning to nest with a moth in her beak to feed her hungry brood. The hen builds the nest and does all the incubating, but the male helps to feed the young once they are fledged.

Dedicated bird parents

We have seen how the male seahorse takes on the sole responsibility for launching his young on the world, and how male sticklebacks take on the care of their offspring unaided by the females. We know that in mouth-brooding fishes male and female may share the task. In birds there are a few species, such as the emu of Australia and the rhea of South America, in which the male does all the work. In a number of other birds the two parents share in the upbringing of their babies. But on the whole, it is the female that assumes most, if not all, of the responsibility.

Baby birds are of two kinds. There are those that are hatched blind, naked and almost helpless, unable to do much more than stretch up the neck and gape to receive food. These are the ones that are called nestlings, and they are referred to as being altricial. The second kind are more developed and are called precocial; they are exemplified by the hatchlings of the domestic hen and the duck and a general term for them is chicks. They have their eyes open when hatched, the body is covered with down feathers and they can run about almost from the moment when they leave the shell. Being able to move about on their own, they leave the nest. They are also able to feed themselves from the first

A pair of mute swans with their precocial cygnets. Swans are devoted parents but once the cygnets are on the threshold of maturity the parents drive them away to seek a home elsewhere.

moments of active life, consequently the role and importance of the parents are strictly limited.

The chicks of the domestic hen have a built-in tendency to peck at any small rounded object that lies in their path. It may be edible or inedible. The chick tries all such objects and so learns what is food and what is not food. The mother helps by calling her chicks, when she sees a particle of food, and indicating to them, by pointing her beak at it, where the food is to be found.

Communication between hen and chick is largely by sound. Even before hatching a chick starts cheeping and continues to do so incessantly until it gets its first coat of real feathers. The hen also keeps up a steady clucking, once her chicks are moving about with her. The value of this two-way communication between hen and chick can be illustrated by discussing two aspects of it.

So long as the chicks are cheeping normally the hen knows they are gathered near her and that there is no cause for alarm on their behalf. Should a chick stray far its cheeping rises in intensity and takes on a more urgent note. On hearing this the hen clucks more insistently, the rest of her brood draw closer to her, and she makes her way with them to the straying chick until it can rejoin the brood.

The experiment has been tried of putting a glass cover over a chick that has been taken away from the brood. Under the cover the chick can be seen to be cheeping desperately, but it cannot be heard. The hen can see it, but because she cannot hear its calls she carries on as if nothing unusual was happening.

The other example shows what happens when the chicks are in danger from enemies. Should a hawk appear in the sky, the hen gives a particular squawk – the hawk-alarm cry. The chicks near her run under her wings and disappear amid her fluffed-out body feathers. Any chicks too far away to reach her quickly freeze, remaining absolutely motionless and crouched to the ground, until the hen gives an all-clear call.

In effect, then, the domestic hen does little more for her chicks than brood them against low night temperatures, shelter them from the rain, show them food and warn them of the approach of enemies.

The parents of 'helpless' nestlings must do more and it may be for this reason that, as a rule, the male helps the female to bring up a family. Nevertheless, they, like the parents of 'developed' chicks, are largely helped by instinctive behaviour in themselves and in their offspring.

Take any one of the small song-birds that nest in our gardens. The four or five nestlings must be fed at frequent intervals, a daunting task even when father assists. The nestlings' food is usually insects or other small animals such as earthworms. Ten thousand trips may have to be made in the search for food before the babies in the nest reach the stage of looking after themselves. Each journey may be from 30 to 300 ft (9 to 90 m) or 150 ft (45 m) on average. So, in the course of two weeks, the usual period for the young to grow in size and grow their feathers, the parents between them fly 180 miles (290 km) or 13 miles (21 km) a day, 26 miles (42 km) if one parent deserts or meets its death. That is only the distance from the nest to the food source; further ground has then to be covered to locate food.

There are other duties for the parents of 'helpless' babies; for example, the nest must be kept tidy and clean. First, the eggshells must be carried away and dropped some distance from the nest – eggshells lying around might betray the position of the nest. A more continuous chore is the removal of the nestlings' body waste, the faeces. This is at first swallowed by the parents, usually the female. Later it is passed by the nestling in a jelly-like sac, the faecal sac, which the parent waits for after feeding a nestling, taking it in the bill and flying away to drop it at a distance from the nest. Dead leaves and anything else falling on the nest must be collected and carried away.

In addition to the regular nest duties, the parents need to keep an eye open for predators, from the air or from the ground and also for changes in the weather. A bird with nestlings may sit tight on the nest for an hour or more under a sunlit sky, and then heavy rain may fall. This means that she has been aware of the approach of rain and has stayed at home to shield her nestlings.

All the time that they are tending their young the parents have to find food for themselves too. The task is made easier by virtue of inherited patterns of behaviour. When a parent carrying food lands on or near the nest, the nestlings feel the vibrations. They respond by opening their beaks wide and by thrusting them directly upwards to have the food pushed into their throats. Once their eyes are open they gape directly at the parent who does not have to remember which nestling was fed last: it is the one that shows the least eagerness to be fed again.

Anyone who has hand-fed an abandoned nestling, and has had to induce it to open its beak to receive food, will appreciate how tiring and time-consuming it is to feed a baby bird that does not co-operate fully. Indeed, the task of the parent birds would be impossible if their nestlings did not react as they do.

ABOVE The emperor penguin chick is kept warm in a fold of the parent's abdomen skin that forms a sort of hood over it.

BELOW Moorhen and chick. This species is remarkable in that the chicks of the first brood help the parents to feed and tend those of the next brood.

BELOW The first action of the cuckoo hatchling is to eject the rightful eggs from the nest, or even the nestlings, which are no smaller than itself. From this inauspicious beginning the baby cuckoo grows enormously, outstripping its foster-parent in size.

From mound-builders to water rails

Our garden bird represents the general picture for rearing 'helpless' young. There is a wider variation in the relationships between 'developed' chicks and their parents. Firstly, there are the chicks that are totally independent from the moment of hatching; these include the chicks of mound-builders. Next come the plovers and their relatives; their chicks follow the parents but find their own food. After them are the game-birds, including the domestic hen, whose chicks follow the parents which show them where food is. Then come the grebes and water rails that follow the parents but need to have the food pushed into their throats. Finally, there are the gulls whose chicks are not fully 'developed', indeed they are partly 'helpless', staying in the nest until well grown and able to walk but nevertheless still depending on being fed by the parents.

To return to the mound-builders, of Australasia – also known as scrubfowl, brush-turkeys, mallee fowl, megapodes (big feet) or incubator birds – they have what is perhaps the most remarkable nesting behaviour of any living bird. Once adult they spend practically all the year building the nest and keeping it at the correct temperature. This is particularly true of the males. Yet, once the chicks hatch, the parents show little or no interest in them.

The mallee fowl, the species most studied, starts to dig a deep pit in June and fills it with vegetable litter. The rains come, the vegetable rubbish heats up and in September the hen starts to lay anything up to thirty-three eggs, each one at an interval of two days or more, in an incubating chamber within the mound.

As soon as the last egg is laid the first starts to hatch. Throughout the months the male, and to some extent the female, is devoted to ensuring that the mound remains as nearly as possible at an even temperature. Using his tongue as a thermometer, the mallee fowl tests the temperature. If the mound is getting too hot the nest is opened up to let some of the heat escape. Should the temperature in the mound fall below the temperature required, the top layer is removed to let in the heat of the sun to warm it.

In contrast to the enormous nests of the mound-builders, the emperor penguins build no nest. In the middle of the Antarctic winter the female lays a solitary egg which she holds on her feet or passes to the male to hold on his feet. This protects the egg from the cold ground and further protection is given by a fold of the parent's skin that encloses it.

After a few days the female waddles away over the ice to the distant sea to feed, leaving the male almost literally to hold the baby. Two months later she returns just as the egg has hatched, fat and with a cropful of food to feed the chick. It is assumed that she judges the time for her return guided by some internal clock.

The best-known example of a bird that builds no nest is the cuckoo. The female lays her eggs in the nests of small birds. Having selected a nest she removes one of the rightful eggs, replacing it with one of her own. When the young cuckoo hatches out it ejects all the eggs or nestlings properly belonging to the nest. The foster-parents are thereby able to give all their attention to this huge foundling.

BELOW Bennett's tree kangaroo of Australia with its baby (or Joey) peeping out of the pouch. One of the most surprising things about the pouch of the female kangeroo is its elasticity. It is hardly noticeable when first occupied by the minute newborn baby. By the time the Joey is ready to leave its mother, the pouch, which is still used as a resting place, has stretched to an enormous bag to contain the growing youngster.

OPPOSITE Giraffe suckling her calf. The baby giraffe is born with eyes open and other senses functional. It is not so vulnerable to flesh-eating predators as most hoofed mammals because the parents themselves can put up a formidable fight with their large hoofs.

Cuckoos have given their name to several insects, like the cuckoo-bees that lay their eggs in the nests of other bees. But cuckoos are not alone in their nest-parasitism. Some weavers and weaver-finches in Africa follow the same practice, so do the honeyguides of Africa and southern Asia. In the New World, the blackheaded duck of South America lays its eggs in the nests of other ducks and the cowbirds, also of America, parasitize the nests of smaller birds.

There is a kind of cuckoo behaviour in an American minnow which lives in the same waters as a green sunfish. When the male sunfish is building his nest the minnows assemble above him. When the female sunfish has laid her eggs the minnows lay theirs in the same nest. It seems that the stimulus which makes the minnows spawn is a chemical (pheromone) given out in the water from the milt or ova of the sunfish. Thus, when a pair of these minnows are in an aquarium with no sunfish present and no sunfish nest, they will spawn if the fluid from sunfish eggs or milt is poured into the water.

The moorhen, or waterhen, of Europe, known in North America as the common gallinule, has 'developed' young, but these are precocious in an unusual sense. The first brood of the year stay with the parents after a second clutch of eggs is laid, and they help in feeding the chicks of this later brood. They also help the parents to rebuild the nest should it become damaged by flood waters.

Studies of the moorhen help to answer the question of how far birds recognize their own chicks. It is known, for example, that a gull will protect its chicks while they are in the nest. Should one stray from the nest before it is time to do so, the parent is quite likely to kill it and eat it. Moorhens do not recognize their chicks as individuals until they are two weeks old.

The 'developed' and 'helpless' young of mammals

The kangaroo is able to carry her young about in a pouch. Thus she keeps it warm and otherwise shelters it, and she can carry it about to new feeding grounds. The placental or true mammals, lacking a pouch, must adopt other measures. The feeding of the young is uniform: they are suckled until they can eat solid food, then they are weaned. Before this, we can see the same two divisions as in birds. There are precocial or 'developed' young, those that are born with eyes open and can walk within a short time of birth, often within an hour. On the other hand there are altricial or 'helpless' young, born blind, naked or with only a covering of short hair, and which at first can only make feeble movements with the limbs; these young are extremely vulnerable to predators and need the protection of a nest.

There is at least as much variation among 'helpless' and 'developed' mammals as there is among the two types of birds, and there are more contrasts. For example, young rabbits are born blind and helpless in a warm nest in a burrow which they leave at about three weeks. They are fully altricial. The grown hare, so like a rabbit in appearance and in its anatomy, has fully precocial young. They are born fully furred, with eyes open and able to walk within a few hours of birth.

The rabbit is born after a gestation of 30 days, the hare after 42 days.

Baby camel walking in step with its mother. It is almost a replica of its parents but with dangling legs and a fine woolly coat. Camels are desert dwellers and nomadic and a baby must be able soon to travel with its mother.

The rabbit's eyes open at seven days. So the main difference between the two is that the extra time the hare remains in the womb allows it to be that much ahead of the rabbit at their respective moments of birth.

Another example of contrast is seen in guinea pigs, which are about the most precocious of mammals. They are born with a full coat of fur and can use their legs to move around almost as soon as the umbilical cord is severed. Their eyes and ears are open at birth, indeed, they are miniature adults in every sense except that they need milk. As for feeding, they start to chew grass at a very early age even although they may not swallow it. Also, they are born with good thermo-regulation: that is, they can keep themselves warm by regulating their temperature, unlike 'helpless' young mammals, which have to be kept warm.

Guinea pigs are rodents and their precocity is in marked contrast to most other rodents, which are 'helpless'. They are born in burrows in a warm nest. They are born naked, with eyes and ears closed, and can only make feeble leg movements. The mother curls herself around her litter to keep them warm and when she must leave the nest her babies huddle together. They have just enough strength in their legs to drag themselves towards their mother, when she is present, or their litter-mates, when she is away.

In general, the larger the species the longer is the period of gestation. The shortest gestation is in mice, rats and shrews, in which it is about three weeks. The duration cannot be given in an absolute number of days because it varies with the individual. In guinea pigs it may be 63 days or as much as 70, in the black rat 21 days or as much as 30.

The longest gestation is in elephants, which also show the greatest variation, from 607 to 760 days, so that the average is around two years. This is well above the rhinoceros, 485–578 days; the giraffe, 426–468 days; the camel, 350–400 days; and the hippopotamus, 225–243 days. For the ass the gestation is around a year and for the horse 264–345 days. Cattle are around 270–290 days. The larger deer gestate for 240–250 days; the smaller, such as roe deer, for 140–290 days.

All those mentioned are herbivores, have 'developed' young and ususally have one young at birth.

Gestation in the carnivores is markedly shorter. Their young are 'helpless' and there are usually two or more in a litter. The longest gestation is in bears, with 187–206 days, but although they are classified with the carnivores they are omnivorous, eating everything, but mainly vegetarian. Among true carnivores, such as the fox, the domestic cat and dogs the period is about 60 days and with the lion and tiger it is just over 100 days.

Human beings can be placed in a category of their own but, for comparison, the human gestation is about 270 days.

ABOVE Like the majority of rodent young, this greater Egyptian gerbil litter is born blind and helpless.

The baby koala leaves the mother's pouch when four to five months old and then rides clinging to its mother.

3
Food and the Pyramid of Life

The ruddy duck of Patagonia lays fourteen eggs which together weigh 3 lb (1·3 kg), this being three times the weight of the duck that laid them. This expresses dramatically the principle, several times referred to in the previous chapters, that in reproduction an animal contributes part of itself to form the next generation. It also reminds us that reproduction not only ensures the continuation of the species: it also supplies food for others. All but one of the new lives in the fourteen eggs laid by each female ruddy duck must, on average, perish before reaching maturity or the world would, in a relatively short space of time, be swamped with ruddy ducks.

Egg-eating snake swallowing an egg. Once swallowed, the shell will be crushed and vomited while the contents are digested. The snake has long spines on the underside of the neck vertebrae which point downward and forward to pierce the egg while special neck muscles crush it.

Food-chains

Food is basic to life. Feeding is the foremost requisite for all living organisms because food supplies the materials for growth and the energy for all activity. The simplest distinction is between those organisms able to use the energy from sunlight to make their own food, and those unable to do so. The former include all green plants and a few single-celled animals. These are known as autotrophs or 'self-nourishers' and all the rest are heterotrophs, literally 'other nourishers', meaning that they feed on other organisms. 'Other nourishers' include all animals, with the few exceptions mentioned. All animals' feeding habits constitute a link in a food-chain.

Some food-chains are simple. A cow eats grass; man drinks milk; he eats milk-products and also the flesh of the cow as beef. Thus there are only three links in this chain, which is simple because the cow is domesticated. If the situation is translated to the wild, however, the chain becomes more complicated. We can do this by substituting for the cow its near relative, the wild buffalo. The buffalo feeds on many kinds of plants. When the female buffalo gives birth to a calf she becomes the centre of attraction for a variety of other animals which gather round to await events. These are called predators, meaning animals that prey on other animals. Vultures in their separate stations high in the sky, invisible to the human eye, are quick to note a recumbent animal. To them it may be a corpse or a quadruped in labour and one of them flying down acts as a signal to others to join. The after-birth may be their only reward.

On the ground hyenas and hunting dogs are drawn by the smell of blood, but a buffalo is not to be treated lightly so they keep their distance. An hour after it is born the calf is walking quite strongly beside its mother. The calf is then possibly even more vulnerable because now the mother's attention is directed more to leading her calf to the herd, and the predators close in. She usually succeeds, however, in keeping predators at bay but if she does not reach the protection of the herd with her infant it is torn apart by the dogs or the hyenas and her long period of pregnancy ends in nothing more than providing food for her tormentors. The sight of such a spectacle excites our pity, yet it is only 'the law of the jungle', and the food-chain in operation.

A buffalo was chosen only because it is the nearest counterpart among wildlife of a domestic cow. Any one of a score or more antelopes or other hoofed animals could have been chosen with equal effect. Either the adults or, more commonly, the young of any of these species are

OPPOSITE The hummingbird is beautifully adapted to feeding on nectar and on very small insects associated with the jungle blossoms. It is able to hover in front of a flower and thrust its long slender beak into the blossom to suck up the nectar and small insects that have crept in to feed on the sweet juices.

In Kenya's Amboseli reserve, near Mount Kilimanjaro, vultures gather to share the remains of a lion's banquet. They spend their time high in the sky out of sight and when one sees a carcass it begins to descend. Its action is seen by other vultures who follow suit.

potential victims of the medium-sized or large animals of prey. Indeed, any animal of whatever species and whatever its size is potential prey to some larger animal, except for the super-predators, such as the lion, tiger and bear, which normally have no enemies other than man.

Buffaloes may be eaten by one of several predators. The buffaloes themselves may graze grass or browse leaves. They probably ingest small animals, although incidentally, with their plant food. So there is no simple, direct food-chain as with the domestic cow. In addition, other hoofed animals may eat the same grass as the buffalo or browse the same shrubs. Also, the vultures, hyenas, hunting dogs and lions take other prey as well, but sometimes take a buffalo calf or an adult.

Food-webs and food-pyramids

Feeding relationships are usually as complex as, or even more complicated than, those represented by the buffalo, and seldom as simple as the chosen example of the domestic cow. This complexity is better thought of as a food-web, a notion expressed by writers earlier in the century who spoke of the web of life. If nothing else, this now outmoded phrase epitomized the present concept of the interdependence or interrelationship of all organisms. The food-web is usually expressed statistically. Another concept, however, referred to as the food-pyramid is expressed quite differently. It is represented by a triangle divided by three horizontal lines. The basal layer represents the 'self-nourishers': the vegetation on land and the plant plankton and seaweeds in the sea. These are consumed by herbivores. Strictly speaking, a herbivore is an animal that feeds on herbs, that is, any soft-stemmed plant, and some authors prefer to speak of phytophages, or plant-eaters, but herbivore is the more commonly used and the better understood.

The next layer up in the food-pyramid contains all the herbivores from the minute shrimp in the sea feeding on microscopic plankton to the huge elephant on land consuming daily very large quantities of grass, leaves, twigs and fruit. Herbivores convert plant material to flesh. They are food for the carnivores and the third layer up in the food-pyramid includes the small carnivores which in turn may be the prey, together with the herbivores, of the larger carnivores that form the fourth layer, or apex of the pyramid.

The larger carnivores are spoken of as the super-predators. They are the large animals that are normally without enemies. Man is a super-predator, as are the lion, tiger, leopard, wolf and hunting dog. They are,

in the sense of the food-pyramid, dominant, although exceptionally one may fall victim to another super-predator, or it may even succumb to the cannibalism of one of its fellows.

Like the food-webs themselves, the names we give to the various types of feeders are prone to be complex and confusing. Thus, a carnivore – strictly speaking a flesh-eater – is an animal that preys upon other animals, even if the flesh eaten is that of such lowly creatures as insects, earthworms or even smaller prey.

A simplified diagram of an ecological pyramid, showing the various stages of the food chain in a savannah-type plant and animal community. The plants or producers are food for the phytophages or plant-eaters, who are in turn eaten by the predators. The latter may themselves become the prey of the super-predators.

Carnivores tend to have weapons of offence in the form of claws, large jaws and sharp teeth, or limbs capable of being used as grasping organs. Some species may have the ability simply to crush their prey, as in the case of pythons and boas. The larger the animal the more effective will be its weapons of offence, hence the existence of the super-predators. Yet even here there is no clear line of demarcation. For example, the elephant is capable of overcoming a tiger or a lion, goring it with its tusks, throwing it bodily through the air with its trunk or trampling it to death with its feet. And an elephant is, by any definition, a herbivore that should be included in the second layer from the bottom in the food-pyramid. The point is, however, that although an elephant may kill other animals it does so defensively, possibly at times in revenge or so it seems. But elephants do not eat their victims so they cannot be classed as carnivores. The same is true of the African buffalo, a herbivore in the strict sense, yet one that is well known as capable of finishing off a lion, and one regarded by human hunters in the African bush as probably the most dangerous of all African animals.

By contrast, and further emphasizing the difficulty of placing animal-feeders into neat and orderly categories, we have the gentle herbivores known as antelopes. They range from the giant eland, the size of an enormous domestic bull to the tiny duikers, little more than 1 ft (30 cm) in height.

The line between a carnivore and a herbivore is often smudged. Lions, dogs and foxes will at times eat fruit or grass, sometimes in large amounts. Equally herbivores will at times eat flesh. In recent years it has become established, by numerous scattered observations, that duikers and other small-to-medium antelopes will, at times, deliberately kill and eat small mammals and ground birds. Some of the best-tasting mutton, so it is claimed, is from sheep that deliberately seek out those pastures heavily infested with snails.

ABOVE Chimpanzees feed on a variety of fruits and nuts. They eat some leaves and young shoots as well as bark, and they will also eat insects and eggs. In a few areas they have been seen to kill and eat birds, young antelopes and snakes, but these carnivorous habits seem to be strictly local.

The pyramid of numbers

The scientific study of food-chains did not begin until the 1920s, when Charles Elton, the English ecologist, noticed that animals at different levels in a chain had definite relationships with each other in regard to numbers and size. He found that herbivores tend to be smaller and more numerous than the predators that preyed on them, and these predators had even bigger, but less numerous, predators feeding on them. Similarly, herbivores are often bigger and less numerous than the plant foods on which they depend.

Elton's studies sparked off intensive studies of food-webs, more especially of those animals that have an economic importance. An outstanding example is the herring, formerly so important as a food-fish in countries bordering the North Atlantic. The 'self-nourishers' at the start of the food-chain leading to the herring are microscopic single-celled plants of the plankton that exist in astronomical numbers in the sea. These are consumed by herbivores, notably small crustaceans. These provide food for invertebrate predators of many kinds and these, together with the 'self-nourishers', are eaten by fishes of which the herring is economically the most important.

OPPOSITE A specialist feeder, the chameleon shoots out its long tongue to capture an insect. A chameleon moves extremely slowly, carefully stalking insects on the foliage among which it lives, and seizes its prey with a lightning-swift thrust of the tongue.

This part of the web uses, however, only about a tenth annually of the plant plankton. The rest of the plants, when they die, sink down through the water to the ocean bed. Some dead material is eaten by various fishes and crustaceans on the way down. Most of it reaches the sea-bed where it decomposes due to the activity of bacteria and gives rise to another food-web, often called the decomposer web. Most of the animals in the decomposer web feed indiscriminately on dead plant material and bacteria. They provide food in the sea for the many predators including bottom-living fishes, like plaice and skate, that are eaten by man.

The simplest method of finding out what food an animal eats is by catching it and examining the contents of its stomach. Some animals chew their food; however, others take semi-liquid or liquid food, all of which becomes unrecognizable or can only be identified with great difficulty, such as picking out the tiny bristles, under a microscope, of marine worms. A quicker and more refined method is to add to the water taken in by the 'self-nourishers', a radioactive substance, but at too low a concentration to damage the plants. This is taken in by the plant tissue and can be traced in the bodies of crustaceans, small fishes and herrings. Moreover, when the plants die the radioactive substance can be traced through the bodies of the animals forming the decomposer web.

There are decomposer webs on land. The carcass of an elephant, weighing five or six tons, can in suitable conditions disappear in two or three weeks. First the hyenas and other carrion-eaters, including vultures, attack the bulk of the flesh. Carrion-eating insects, of which blowflies are an example, and other small invertebrates join in. Rodents clean the bones of remaining shreds of flesh and even gnaw the bones, and hyenas with their powerful jaws and teeth crack the long bones. The skull and ivory tusks may be left but the decaying fragments of flesh, consumed by bacteria, and the droppings of larger decomposers, fertilize the soil so that grass and herbage generally grow vigorously drawing a green mantle over the whole scene.

A simple example of the pyramid of numbers for an acre of ungrazed pastureland has been given by Eugene P. Odum, the American ecologist. The living inhabitants numbered six million 'self-nourishers' producing protein for the next up the chain, representing the lowest level of the pyramid, seven hundred thousand herbivores, three hundred and fifty thousand small carnivores, and three large carnivores.

There are, however, food-webs that do not show such a clear pyramid of numbers. Those in which trees are the autotrophs are examples. In these the succession is, in a sense, inverted. The largest and least numerous, that is, the trees, are at the bottom level, and there are relatively more small herbivores and a large number of carnivores and parasites of a great range of sizes feeding on them. This is only one example of how the conventional pyramid of numbers cannot always be rigorously applied.

If we compare the food-web in the coastal waters of the sea with that of a forest this difficulty becomes even more apparent. The 'self-nourishers' in the sea are almost entirely microscopic plants, especially diatoms. Even a small carnivore, like a tiny crustacean eats hundreds of them a day. Moreover, the diatoms are short-lived so in the course of even its own brief lifetime a crustacean feeds on many generations of diatoms. Trees, on the other hand – one of the 'self-nourishers' on land – are long-lived and they include material that has been accumulating for years, in sharp contrast to the short life of the plant plankton.

African genet, a small cat-like animal related to mongooses, which feeds largely on birds caught in the trees and bushes. It is mainly nocturnal and its diet in the wild has not been fully studied. In captivity, the genet feeds on meat and will readily eat large night-flying beetles, but its ready response to the chirping of a young bird clearly indicates what must be its main quarry under natural conditions.

The pyramid of biomass

To get over this difficulty a pyramid of biomass is used. The biomass is the total dry weight of any particular species measured at a given moment in a given area. The reason for taking the dry weight is that different animals contain proportionally more water than others.

It used to be said that the European mole ate more than its own weight in food a day. When, some twenty years ago, a careful study was made, it was found that a more reliable figure was 60% of the mole's weight. The mole's food is mainly earthworms but it also eats insects, millipedes, slugs, and even some plant food. A mole that is feeding wholly on earthworms must consume a larger weight of food than one eating mainly insects. This is because an earthworm's tissues contain a high proportion of water, while an insect's body contains a higher ratio of protein and is therefore lighter.

Omnivores, the all-eaters

Man, whatever his race, colour or creed, tends to represent human virtues and vices by reference to animals. A person with an insatiable appetite is usually compared with a pig. If, however, the domestic hog over-eats, it is largely because it has nothing else to do, and it is encouraged to over-eat because that is the surest way of producing a good return in terms of pork.

It has also been said of pigs that they will eat anything but broken glass. This is virtually true of domestic pigs which will eat anything animal or vegetable; they will even chew wood or eat carrion. They will happily chew cinders from a coal fire. They are, in fact, perfect omnivores.

A cheetah with a Thomson's gazelle which it has outrun at a speed of 45–60 m.p.h. (72–96 km/h). The cheetah is usually referred to as the fastest land animal, commonly credited with speeds of no less than 70 m.p.h. (112 km/h). Such speeds may be attained but there is reason to believe that the slower speeds are the more usual.

A wild pig will eat roots, bulbs, tubers, leaves, fruit and toadstools. It also takes snails, earthworms, reptiles, birds, small mammals such as rats, and any eggs it can find, as well as any dead meat it may come across. Contrary to popular belief, it never over-eats.

A glutton is one who over-eats. The glutton is the largest member of the weasel family and is also called a wolverine. There are many legends about its voracity. In real life it eats mainly carrion; will kill a sheep, deer or even a moose; will rob any trap; drive bears or pumas from their kills; kill and eat porcupines, and drag a carcass three times its own weight for a mile over rough ground. In early summer a wolverine will eat ground-nesting birds, in late summer it takes wasp grubs, in winter berries. And it will chew leather. Legend is unnecessary. Yet, for all this, the wolverine never over-eats.

Predators, the benefactors

So much talk of killing led, in the past, to the famous saying about 'Nature red in tooth and claw', yet the truth, as is now being realized, is that predators benefit the species they prey upon because they take mainly the old, the diseased and the weak. Moreover, the great majority of victims are despatched painlessly.

One result of the intensive field studies that have been carried out since the middle of this century is a realization that the dice are not so loaded against prey species as was previously supposed. On average, it now transpires, predators are successful in making a kill in only about one in every three attacks made and it is mostly the weaklings, the sick, the elderly and the young that are eliminated.

Mention has been made of the high mortality each year among the young. It is only possible to guess at the causes using scattered observations as a guide. Undoubtedly deaths due to a wide variety of accidents, often the result of inexperience, account for a high proportion of the total. Congenital defects, disease and lack of food also play a part. Deaths from fighting represent a very small fraction of the whole and deaths from predators are not a great deal higher. This was brought out strikingly by a survey carried out some years ago on the isolated island of Skokholm off the coast of Wales.

There was a population of about two thousand rabbits on Skokholm but no predators. During the breeding season the population rose to twenty thousand, yet by the beginning of the next breeding season the numbers fell back to two thousand. Although it was not possible to analyse precisely the causes of death they must have included accidents, parental neglect, congenital weaknesses, disease and old age. The survey was made before the introduction of myxomatosis so the results could be typical for all animal species in a predator-free environment.

Energy flow

Food-chains, food-webs, food-pyramids, pyramids of numbers and of biomass may be relatively uninteresting subjects except to the specialist. Yet an understanding of them is essential to our comprehension of what is happening within the family of animals. It is important also in matters of conservation or even of the future of the human race.

Chains, webs and pyramids are, in a way, little more than fumbling

The blackbacked or saddleback jackal is a scavenger of the African grasslands where it feeds largely on carrion. It also kills hares, rodents, birds such as guinea fowl, reptiles including pythons, and eats insects, eggs, fruits and berries. Blackbacked jackals live in pairs or small parties.

attempts to represent, on paper and in a simple form, immensely intricate situations. Each has its limitations and its drawbacks and ecologists are turning more and more to energy flow as the most accurate measure of what is happening. By energy flow we mean the amount of energy contained in the food eaten by animals. It is usually measured over a period of a year.

Study of energy flow has revealed enormous losses in energy. Thus, when a cow grazes, the food it takes in contains stored energy derived from the sun. Some of this energy is used by the cow in making its body work. More energy is lost in the form of heat radiating from its body into the surrounding air, and only about a tenth of the energy from its food is used in growth of its body tissues.

Suppose a cow dies and its flesh is eaten by foxes: 90% of the energy contained in the beef goes in making the fox's body work – that is, in what is called its physiology – and in heat radiated from its body. If the fox is eaten by a lion, there is a similar loss of 90% of energy.

The ecosystem

Food-chains and webs and the different kinds of pyramids make up what are known as ecosystems. The name was first used in 1935 but only in recent years has it become at all familiar. An ecosystem consists of a habitat and the community of plants and animals inhabiting it. The two components are inseparably linked, for if the plants and animals were taken away there would be no habitat.

Ecosystems are of many kinds and many sizes. A small pond is an

The red fox has secured a rabbit, its main prey before myxomatosis struck. Its diet is a wide one, however, and includes any small rodent or bird, carrion, and berries in season.

ecosystem, so is an ocean. A field of grass is an ecosystem, so is a steppe stretching in all directions to the distant horizon. All are separate yet none is wholly independent for animals may migrate from one to the other, plants may be blown by the wind and soil carried by flood waters. Nevertheless, the ecosystem is a useful concept within which can be expressed certain basic cycles of life, because although they vary in size ecosystems share certain principles and processes.

The green plants take in carbon dioxide from the air, absorb energy from the sun and take in minerals from the soil to form organic compounds which provide food for the whole ecosystem. So they are called the 'producers' for the whole system. The rest, which include bacteria, fungi and animals obtain their supplies of energy and mineral nutrients from the green plants or from each other. The nutrients are taken into the bodies of animals, which are known as the 'consumers' and when these die their dead bodies are eaten by other animals, as well as bacteria and fungi, which are classed as 'decomposers'. Thus, an antelope eats plants, a hunting dog eats the antelope and when the dog dies its decomposing body is eaten by smaller animals and bacteria and eventually all of it is returned to the earth, so there is a constant energy flow and a circulation of nutrients throughout the ecosystem.

In an aquatic system the minerals are dissolved in the water, so are the gases (oxygen and carbon dioxide) from the atmosphere, and the plants are mostly smaller than those on land. Otherwise the energy flow and the nutrient circulation are in principle the same as on land.

The importance of plant food

In Tanzania's famous Serengeti National Park large concentrations of hoofed game can exist in relatively small areas of grassland. A million head of game can live on some 770 sq miles (2000 sq km) of grassland interspersed with acacia woodlands and forest with rivers. The biomass, the weight of animals to a given area, is probably greater there than anywhere in the world. This is because the different species of game

ABOVE The Nile crocodile and the fish-eating goliath heron share the same habitat but not the same food. When young the crocodile is a predator on insects and fishes and when fully adult is a super-predator.

ABOVE Giraffe's head up among the highermost branches where no other terrestrial animal can reach.

BELOW Group of kudu, one of the larger African antelopes, at a waterhole, with an old male and a young female drinking.

animals are separated ecologically, some by their food requirements and others by their different preferences in habitat.

There are three animals that can be compared to help explain this high biomass: the giraffe, the gerenuk and a gazelle. The giraffe with its long neck can reach up to midway of a tall tree, or the tops of any tree less than 18 ft (5·5 m) high, to browse the foliage. The gerenuk, a long-necked antelope, will stand on its hindlegs and reach up to take foliage at heights below those normally browsed by giraffes. Any one of several gazelles, with necks of normal length and without the habit of standing erect on the hindlegs, browse the foliage even lower.

Another grouping of a different kind is that of zebra, wildebeest, topi and Thomson's gazelle. All four are grazers and may feed off the same grass plant, yet do not compete for their food, as each uses the plant at a different stage in its growth. A zebra eats the outer part of the stem, which is too tough for antelopes to chew and is not sufficiently nutritious for them. These parts are acceptable to the zebra because it has incisors in both upper and lower jaw for cutting through the wiry stems, and its digestive system is able to cope with a high throughput rate of low-quality food. The topi has a pointed muzzle enabling it to reach the lower parts of the stems, which the other grazers cannot reach and the wildebeest with its square-ended muzzle can pick off the horizontal leaves. Within a few days of all these three animals having clipped them, the grass tufts sprout from the base and Thomson's gazelle with its small muzzle is able to nibble the new growth.

Although the grass is so heavily grazed by this combination the pasture still does not suffer; if anything it benefits more than by being used by a single species. This combined grazing merely means that there is less waste by the various parts of the grass tufts dying off and rotting away. The topi is of additional benefit as a grazer because it can eat the drier grass stalks. And if we enlarge the picture to include elephants, we find that they are beneficial because they can reduce extensive areas of long tough grasses to short grass which other animals can then use.

An ever-present problem today is that of feeding an ever-increasing human population in a world in which the area of land available for food-production is decreasing yearly.

There are many who have adopted vegetarianism, rejecting the flesh of animals as an article of diet. Some have done so on moral or humanitarian grounds, others because they believe such a diet to be more health-giving. One result of this practice is to leave no doubt that the human species can achieve health, a capacity for hard physical labour, and a long life without ever eating animal flesh.

Ecologists studying the future of land-use, in relation to the exploding human population, are being led to the conclusion that raising stock for their meat is the least efficient method of food-production. The inevitable corollary is that either there must be great improvements in the method of stock-raising or a greater reliance on vegetable proteins. Plants supply ten times the amount of protein per unit of land surface that animals do, and this is only partly offset by animal flesh being the more efficient way of supplying protein because it is both more easily digested and richer in protein.

4

Escape and Flight

The three primary needs of any animal are food, a mate in order to reproduce and the ability to escape from enemies, its enemies in this context being predators although it could be endangered by adverse conditions such as weather. The first two of these needs have already been discussed and here we deal with the third.

Camouflage

Any good naturalist will tell us – indeed, as will any keen observer – that normally any wild animal is difficult to see so long as it remains still. The moment it moves it begins to give away its position, no matter how well it is camouflaged.

It follows, therefore, that camouflage, and there are many remarkable examples of animals indistinguishable from their backgrounds, is effective only when the animal is not moving. Many animals also use immobility as a first defence when alarmed. The human stalker learns to approach his quarry slowly, watching it closely, and whenever the animal turns its head in his direction he remains absolutely still. Animal hunters, such as cats, which stalk their prey use the same tactics quite instinctively.

The prey species uses similar tactics. A common trick is for the intended prey to remain immobile until the predator is almost within striking distance, then suddenly to take to its heels or fly away, whichever method of escape is being used.

In the stalking and in the escape from the stalker, conspicuous colours are a disadvantage, for obvious reasons, but neutral or inconspicuous colours are a great help. The stalker especially may benefit from having spots and stripes because these break up the outline of the body, so rendering the stealthy approach even more effective.

Speed on land

Another adjunct in escape is the ability to move quickly, especially to have speed in the initial acceleration as well as while running.

In warfare the introduction of a new weapon often leads to the invention of a counter-weapon which, in turn, brings about the development of a more refined device by the opposing side. This search for continual improvement bears witness to the see-saw tussle of wits between opposing military brains.

It is supposed that this same kind of see-saw effect has been at work in the animal kingdom. There is no direct proof of this since major changes in animal structure take place over longer periods of time than are covered by the period of human history let alone the lifetime of any one observer. The deductions we can make, however, do appear reasonable.

Thus, the study of selected fossils of ages varying up to fifty million years suggest how the present-day horse evolved. Whether the conclusions about the evolution of the horse are valid in every detail or not is unimportant. The fossils furnish a picture of what might have happened and why. More importantly, this picture can be applied to other animals, serving, therefore, as a yardstick to aid our understanding.

The earliest ancestor of the horse was little larger than a hare, with four relatively short legs and four toes, presumably each bearing a claw, on

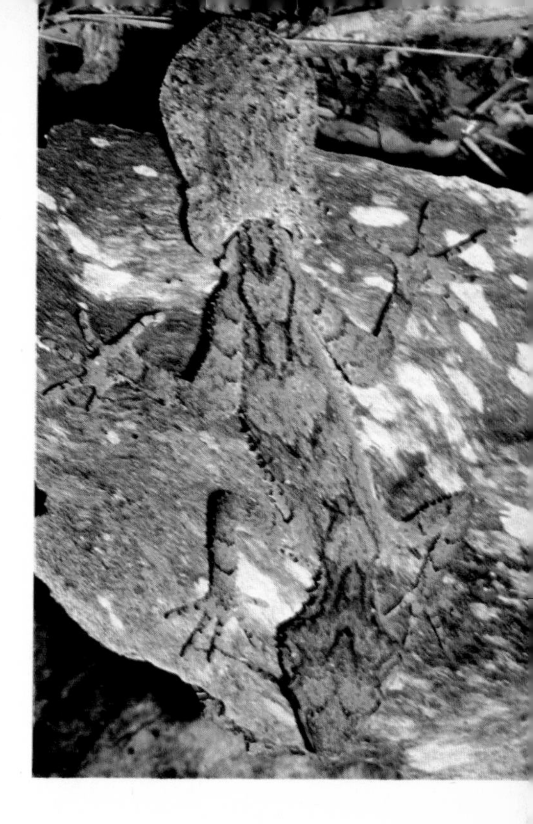

ABOVE By its coloration as well as its shape the leaf-tailed gecko blends closely with its background.

ABOVE The red kangaroo escapes its enemies by tremendous leaps of up to 30 ft (9 m) long and 10 ft (3 m) in height.

OPPOSITE A group of zebra making a quick getaway from a pack of Cape hunting dogs.

each foot. The skull and teeth, as shown by the fossils, resemble closely those of the present-day horse.

In slightly younger geological formations have been found further skeletons of slightly larger horses. As the millions of years passed the body of the horse changed little in its proportions although the size increased. The main changes were in the shape of the teeth, which indicated progression to a complete grass-eater. The relative length of the legs changed little except for the foot. In this, first the outer toe disappeared completely, leading to a three-toed condition, then the middle toe became longer and stronger while the toe at either side of it became weaker finally disappearing except for vestiges, which in the modern horse are known as the splint bones.

From this series of fossils we can imagine the ancestral horse living in forest or scrub, feeding on leaves, and capable of moderate speeds. Then, either because they migrated to grasslands or, more probably, because of climatic changes causing major alterations in the vegetation, the descendants of the horse became grass-eaters in a more open environment and probably found themselves beset by faster predators than their ancestors had known. Speed became important.

ABOVE The pronghorn, the fastest land animal in North America, its speeds rivalling those of the cheetah. This did not save it from men's weapons. The pronghorn was a favourite target of the hunter and formerly numbered something like 40 million. Then there was a severe reduction at the end of the 19th century due to persecution and in 1935 there were only 30,000 left. Now the pronghorn is conserved and its numbers are back to something like half a million.

LEFT The flightless ostrich takes refuge in running and with its long powerful legs can overtake and then outstrip even a fast antelope.

OPPOSITE This three-toed sloth, like all sloths, moves very little and it finds security in moving very slowly.

To achieve greater speed the succeeding generations of horses ran more and more on the middle toe. Those animals born with slightly longer middle toes found themselves with a greater turn of speed and more chance of surviving when pursued. So the middle toe increased in length, at the expense of the others, and the horse's claw became increasingly larger and developed into a hoof.

There are more ways than one of increasing speed. One of the fastest animals, certainly the fastest of hoofed animals, is the pronghorn of North America. It is credited with speeds of up to 60 m.p.h. (96 km/h). The moment it is alarmed it dashes off and no other animal on the American continent has any hope of catching it. Nor has the hunter, even with a rifle, much hope of shooting it. A pronghorn has cloven hoofs, so the four single hoofs of the horse jointly provide only one means of achieving speed.

There is no indication among the known animals of North America that the pronghorn's ancestors developed their excessive speed as an escape from enemies. There are instances among other animals suggesting that changes in structure sometimes come about by chance and not in response to a need. Whether this was so in the case of the pronghorn is a matter of speculation.

What we do know is that the pronghorn has an overweening sense of curiosity, which is made use of by hunters. Waving a white flag, for

instance, causes the animal to draw near to inspect this apparently strange object, bringing it within gunshot. The pronghorn may need its speed to offset this intense sense of curiosity.

The fastest land animal is the cheetah of Africa and southern Asia. It has been credited with speeds of up to 70 m.p.h. (112 km/h), although it is doubtful whether such speeds are commonly used. It also has remarkable acceleration and can reach top speed in a matter of yards. However, it cannot maintain its top speed for more than a quarter of a mile (400 m). Despite having feet much like those of any other member of the cat family, the cheetah's speed is achieved mainly by having unusually long legs in relation to its lithe body.

ABOVE LEFT The roe deer, one of the smaller deer, normally keeps out of sight in the undergrowth. When in the open it runs rapidly and can take enormous leaps for its size. Although only a little over 2 ft (61 cm) high at the shoulder it can clear a fence 6 ft (1.8 m) high or leap 21 ft (6.4 m) in the horizontal.

ABOVE Dolphins are streamlined, built for speed through water. Usually they swim straight ahead but at times break the surface, performing leaps in a movement known as 'porpoising'. This may indicate high spirits, a form of play, but it may also increase the speed at which the dolphins travel.

Speed in the sea

The attainment of high speeds is only one of the many ways in which the higher animals have outstripped the lower. Invertebrates generally lead successful lives either fixed to one spot or else moving about sluggishly – except for one branch of the molluscs, namely the squids, and many of the insects. It is not yet clear why the squids and many insects move comparatively rapidly.

Squids illustrate an entirely different form of locomotion, which is jet-propulsion. A relative of the better-known octopus, a squid has a torpedo-shaped body and ten sucker-covered arms compared with the eight arms of the octopus. Inside a pouch on the underside of its body it has gills. In breathing it draws in water through a tubular siphon to bathe the gills and pumps it out again. The effect is much the same as in fishes which gulp water, allowing it to pass across the gills.

In moving from place to place a squid changes its rhythm of breathing. Instead of a gentle sucking in and squirting out of water it performs the same action more vigorously. If a squid is in a hurry it

increases the speed of its siphon. The outgoing jet meets the resistance of the water surrounding the body driving the squid backwards.

In human aviation jet-propulsion marks a zenith in travel through the air, whether of an aeroplane or a rocket. However, it is only exceptionally used by animals. Over and above the squid, octopus, cuttlefish and nautilus, all jet-propelled, only a few other instances are known. Scallops normally lie on the sea-bed but if their enemy, a starfish, appears they rapidly close their shells driving out water, and then continue clapping their shells. It is a crude form of jet-propulsion but it lifts the scallop off the sea-bed and carries it backwards in a swimming movement. Certain aquatic insects also use a simple jet-propulsion, taking in water, as they normally would in breathing, but driving it out forcibly through the hind-end of the body.

Of the invertebrates only the squid achieves any notable speed and even this is well exceeded by the more usual swimming of aquatic animals, which is a serpentine wriggling of the body, with the tail muscles supplying the motive force. The most spectacular speeds are attained by the sailfishes. They are said to reach speeds of 50–60 m.p.h. (80–96 km/h).

Such excessive speeds can hardly have been evolved for escape from enemies. Sailfishes feed on other fishes, but their prey, fast-moving compared with the general run of fishes, reach nothing like these speeds. One can only speculate that high speeds are useful for carrying sailfishes from one feeding ground to another or for carrying out long migrations. Yet there remains the sneaking suspicion that, as in human aviation and

RIGHT Gentoo penguins swimming. Penguins are flightless but are master swimmers, using their wings as paddles. In effect they fly under water but on land are completely earthbound and rely on their strong short legs to carry them with a waddling gait from place to place.

The kestrel, known in North America as the sparrowhawk, by rapid rhythmic beating with its long wings, can hover almost motionless in the air to survey the ground below. Hovering, diving to snatch prey from the ground, then rising and hovering again is a most unusual method of hunting.

other forms of travel, there is present what seems to be an urge to use speed for its own sake.

Speed in the air

It is in the air more especially that high speeds are attained and flight has been developed as a means of travel most often throughout the animal kingdom. Speeds of 100 m.p.h. (160 km/h) have been recorded for the golden eagle and there may be other eagles which can exceed this. A falcon stooping on its prey may notch 120 m.p.h. (193 km/h).

Certainly it is within the realm of animal aviation that the most spectacular feats of locomotion have been recorded. The most surprising are perhaps those of the common swift. Legend dating from medieval times had it that the swift roosted in the heavens. For once the legend was correct. It has now been established, thanks to the development of the aeroplane, that swifts rest on the wing high in the sky, possibly taking brief naps. Once that became known, and interest focused on the subject, it was realized that a swift never voluntarily comes to the ground. It nests high up in cliffs and buildings and it is only during the breeding season that it is not airborne. A survey in the last few years of swifts in Africa, the bird's winter quarters, has led to the conclusion that there they never perch nor come to ground at all.

It has been said that should a swift become grounded it cannot take off again. In fact, it can do so by scrabbling laboriously up the trunk of a tree, using claws and flapping wings, and out along a branch. Then, having gained sufficient height, it drops from the branch, spreads its wings and flies away at up to 60 m.p.h. (96 km/h).

Another super-aeronaut is the albatross, especially the wandering albatross, with its long narrow wings giving a span of nearly 12 ft (3·9 m). Albatrosses breed on oceanic islands and breeding time is the only time they come to land. Except for those occasions, the wandering

albatross circles the world gliding over the seven seas, using only occasional wingbeats, depending most of its time on updraughts from the waves to keep it airborne. Its method of flying must be the most economical in the animate world.

Flamingos spend much of their time in massed companies sifting food from the water with their curious bills. When disturbed they spread their wings and, with long strides through the water to gain momentum and lift, they take to the air in a body in one of the most spectacular exercises in flying to be seen anywhere.

Powered flight, using forelimbs converted to wings, and the massive breast muscles, make it possible for birds to travel long distances in relatively short spaces of time. Thus, birds can escape from one, if not two, natural enemies other than predators: inclement weather and shortage of food. Swallows, small birds by any standard, can winter in the southern hemisphere and breed in the northern, making a trip of 2000 miles (3200 km) each way.

Other astonishing migration flights are known, like those of the ruby-throated hummingbird, far smaller than a swallow, that spends the winter in tropical America and the summer in Canada. The American golden plover flies from its summer breeding ground in Alaska, to Hawaii, a sustained flight of 6000 miles (9650 km) over the sea with no chance of landfall.

These and other similar spectacular migration flights are overshadowed by that of the Arctic tern. This breeds in the Arctic and then flies south over the Atlantic Ocean to sub-Antarctic islands 11,000 miles (17,700 km) distant. Not least of the marvels in the tern's annual round trip of 22,000 miles (35,400 km) is that it finds the small islands of its destination with no landmarks to guide it.

Other flying animals

There are only three other types of animals, other than birds, that use powered flight. These are insects, bats and the freshwater flying fish. There are many more animals that glide and there are so-called 'flying' types of fishes, squirrels, phalangers, lemurs, snakes, lizards, frogs and many others. Except for the flying fishes, all live in trees and use gliding membranes of some sort to carry them easily from one tree to another in their search for food.

The least accomplished is the flying snake. This launches itself into the air, holding its body rigid, and cleaves the air like a dart. The flying frog has large webbed feet, the webs acting as parachutes. Most others have a membrane on either side of the body, stretching between forelimbs and hindlimbs. Launching itself into the air, the gliding animal spreadeagles its legs, so stretching its flank membranes to the full. By this means it can travel from one tree to another up to a hundred yards away. It loses height in doing so, but instead of having to descend one tree, run a hundred yards and then climb another tree, it merely has to climb a short way up the second tree to launch itself again on another hundred-yard trip. Short of using powered flight, it is difficult to imagine a greater economy of effort.

When we speak of flying fishes we normally mean the marine fishes with elongate breast fins that swim rapidly to the surface, launch themselves into the air, spread their fins to become airborne and glide up to 450 ft (137 m), at a height of 20 ft (6·5 m) or more, before dropping back into the water. The four-winged flying fishes have long pelvic fins as well as elongated breast fins.

It is always assumed that all flying fishes 'fly' to escape enemies, the main one being the dolphin fish. It has also been reported, by several observers, that at the end of the glide, when the fishes drop back into the water, the dolphin has kept pace with their flight and is waiting to snap them up. If such observations are correct, then the gliding of flying fishes seems an ineffectual way of seeking to escape from an enemy.

Flying fishes are related to garfish, often called the long-toms, which have torpedo-shaped bodies and the snout drawn out into a beak or cutwater. Long-toms seem to be playful, often leaping out of the water and somersaulting in mid-air or somersaulting over a turtle at the surface or over a floating log. Perhaps flying fishes are indulging in *joie de vivre* rather than escape from enemies!

There are other fishes with even larger pectoral fins than the flying fishes. The flying gurnard is one. Yet there is no evidence that the fins are

The greater glider, one of the flying phalangers of Australia, is 3 ft (1 m) or more long. It lives in trees and to search for food it launches itself into the air and spreads its legs sideways, stretching the gliding membrane that reaches from the front legs to the hind legs. Using this natural parachute the greater glider travels from tree to tree. It has been recorded as travelling 1770 ft (739 m) in six successive glides.

Gibbons are the smallest of the tailless apes. Their arms are noticeably longer than their legs, and when walking on land they usually hold them above their heads. Once in the trees, their natural element, they use these long arms to swing from branch to branch with superb grace. This series illustrates a gibbon's 'brachiating', as the arm swinging locomotion is called.

used for flying either through the air or under water. Other fishes having large pectoral fins – like the skates and rays, and the largest of them, the mantas – swim with the graceful movements more associated with large birds flying through the air. Mantas also take leaps into the air, as if urged by some impulse to emulate the flying fishes. Yet in no sense can their acrobatics be likened to aerobatics.

The freshwater flying fish of West Africa can achieve powered flight but only over short distances. It is a deep-bodied fish with enlarged breast muscles, as in birds. When it becomes airborne the enlarged breast fins, which are the equivalent of forelimbs and therefore comparable with birds' wings, are vibrated rapidly to give a sound similar to a large airborne dragonfly.

Some spiders use a method of gliding that amounts to parachuting, and therefore merits a mention. Only the young spiders use this form of transportation. They climb to the tops of tall grasses or posts, direct upwards the tip of the abdomen, with its spinnerets (spinning organs), and let out strands of silk which are caught by the wind. On a warm day in summer, with a gentle breeze, thousands of these tufts of silk can be seen floating long distances through the air, the spiderlings on them making landfall one by one.

The aerial travels of the spiderlings are important for this is a method of spreading the population, and that may well have been one of the main reasons for the development of flight in insects.

The smallest flyers

The first insects, we know from fossils, were flightless. There are many primitive insects today that are flightless like the silverfish, the silvery 'living fossil' that feeds on flour and scuttles across the kitchen floor when disturbed. There are others that are secondarily flightless, that is, their ancestors had wings but these have been lost, as in bed bugs and fleas. The latter rival the kangaroo in taking enormous leaps, thus possibly saving energy in their locomotion.

Insect wings originate as flaps of the horny covering of the thorax. These are moved by the action of muscles which alter the shape of the thorax and this moves the wings so that the insect is able to fly with a minimum of effort.

Flight in insects achieves a variety of purposes. It can be used for migrations, for finding a mate, for seeking food (notably in locusts) and for escape from enemies, just as in birds and bats. It is also used for spreading the population, just as any other form of locomotion is used in other animals, to prevent overcrowding but there is a special difference from other animals here.

The most famous long-distance insect migrations that are regular and seasonal are those of the monarch butterfly that spends the summer in southern Canada and winters as far south as Mexico. Its annual migrations are a tourist attraction in southern California, where it rests overnight in large numbers in the trees. Other long flights of insects have been known but were regarded as purely local phenomena until the late 1930s when Captain T. Dannreuther, in England, began to collect records. Now it is known that there are many insects, especially butterflies, that make long-distance migrations, in direct flights, from northern Africa northwards across Europe. The reason for these is as yet not clear and much more study will be needed to elucidate them.

Insect flight as used for seeking a mate finds its best expression in some of the larger moths. Males home on the perfume given out by females and may be attracted to them over distances of a mile (1·6 km) or more.

These flights, as well as those made in search of food and the escape from enemies, represent a logical use of wings once these had been acquired. Flightless insects equally migrate, mate, find food and escape from enemies using ordinary locomotion. It would be fruitless to try discussing here why certain insects first grew membranous flaps on their thorax which in later generations became wings. What we can examine is the advantage to these insects in having wings at all.

Most invertebrates have a life-history that starts with the fertilized egg and passes through one or more larval stages before achieving the adult stage. Insects are no exception although in many there is a pupal stage intervening between the larva and the adult. In all insects, the larval stage is the longest in time and the most important in the life-history because it is the period of maximum feeding and growth. In many species the adult does not feed, and in all there is no growth in size once the adult stage is reached; in fact the adult stage is merely an appendage tacked on for purposes of reproduction.

The situation is underlined by those insects in which the larval life lasts two or three years while the adult lives only a few hours or days, usually no more than weeks at most. At the one extreme we have the mayflies, the wingless larvae of which spend two or more years at the bottom of the river, feeding and growing, which change to adults that immediately

mate and then die within hours. The typical mayfly has been given the appropriate scientific name *Ephemera*, to mark the ephemeral nature of the adult. Yet, by contrast with some other insects, in which the total life-history covers no more than a few weeks, the mayfly is long-lived, if its larval life is included.

At the other extreme can be placed the seventeen-year locust of North America. In this, the eggs are laid in the earth and the larvae hatching from them live for seventeen years, feeding and growing. Finally, at the end of this long period, the larvae become adult, emerge from the soil and enjoy a brief season during which they mate and the females lay their eggs in the ground to give rise to the next generation.

Insects are spectacularly the most successful group of animals, with the largest number of known species, about 90% of the total of all animal species living today. They are the most numerous and the most diverse in structure, behaviour and feeding habits. This incredible success must owe much to the way flight has been used in the adult for providing the maximum scope for expansion and spread of the species at the critical stage in its life-history.

A pipistrelle bat hawking insects. It is the original flittermouse, 1½ in (38 mm) long in head and body with a wingspan of 8½ in (22 cm). It comes out at dusk and hunts small insects in a fluttering flight about 12 ft (3.6 m) from the ground.

5
Comfort and Sleep

Since the late 1920s, a few zoologists in Europe have begun to take a fresh look at animal behaviour. They recognized that the basic needs of animals are, as we have already said, food, a mate and the ability to escape from enemies. They have, quite rightly, added a fourth: the need for rest. They also noted that there were many, usually trivial, things that animals do that are outside these four requirements. They called them comfort movements. These include many quite trifling actions and movements, seemingly without relevance and were difficult to explain until it was realized that they are carried out to provide comfort.

Chafing

One of the first studies of comfort movements was made by two Dutch scientists on certain freshwater fishes of the African lakes. The scientists listed what they saw and the first movement they called chafing. In this the fish suddenly dives to the bottom, lies on its side and rubs its flanks on the sand, on a stone or even on another fish. Or it may keep the vertical position and rub its belly on the bottom.

The action recalls a whole host of similar actions: a cow rubbing itself against a post; a bear rubbing its rump against the trunk of a tree; or a horse, donkey or dog rolling luxuriously on the ground. Some people, when stripped to put on night-clothes before going to bed, take a rough towel to chafe their backs. They take the ends of the folded towel in each hand and pass it vigorously backwards and forwards either diagonally across the shoulder blades or horizontally across the back. It is a simple form of massage and it induces a feeling of well-being that lasts through the night, enabling the person to awake the following morning fresh and ready to face the labours of the day with equanimity.

Zebra foal enjoying the luxury of rolling in the way all members of the horse family do. Rolling removes loose hairs, disturbs skin parasites and gives the animal itself a pleasing sensation in the nerve endings of the skin.

There was one gentleman who had a rough bathmat beside his bed and his last action before slipping between the sheets was to stand erect on the mat and rub the soles of his feet backwards and forwards on the mat, as if rubbing the mud from his shoes on the doormat. An eccentric of the late nineteenth century used to pay young boys a very small sum of money to tickle the soles of his feet, as he lay in bed, with a stout feather. The effect of this may have been to draw the blood to the feet; this might induce relaxed sleep.

One could no doubt list many more such aspects of behaviour on the part of human beings, and indeed of animals, and they would all have one thing in common: they stimulate the flow of blood through the surface capillaries (minute vessels) of the skin. When the blood is made to flow more rapidly at one point on the body, the movement must be communicated to the rest of the blood in the circulatory system. The red corpuscles, carrying food and oxygen, are thus encouraged to carry their precious loads to where they can more effectively benefit the tissues. The white corpuscles, with their capacity for ingesting bacteria, and waste substances, carry out their task more efficiently, to the benefit of the health of the body as a whole.

OPPOSITE Spiny mice basking together on a stone in Kenya. Sunlight must be important since even owls and badgers, nocturnal animals, sometimes bask.

Any simple massage of this kind must be beneficial, if only to a slight degree. Because civilized man wears clothes he fails to carry out simple forms of massage at random intervals during the day and is the victim of a sluggish blood flow. Exercise supplies a remedy, so does deep

Young lions with full bellies luxuriously asleep on a branch. All members of the cat family seem to spend a large part of their time resting if not fully asleep.

breathing. So does the wind on the naked skin, which may be one reason why some people practise naturism!

Man usually indulges his eccentricities in private. Animals cannot enjoy this privilege, so we are more likely to see their actions and be puzzled by them, forgetting that we, too, are part of the family of animals and have similar needs and desires.

Rolling

We once had a donkey called Sally. Like all members of the horse family, she would suddenly sink to her knees and roll on to her back. Having done so she would roll back and forth, then get back to her feet, shake herself, giving off a cloud of dust, and resume what she had been doing. Sally loved nothing better than to roll in the ashes of a bonfire while they were still warm. No doubt this was because she enjoyed the warmth as well as the tingling effect of the abrasive ashes.

People often describe how their dog, usually a male, delights in rolling

in muck and then coming home stinking. When they know that you are a zoologist they ask the inevitable question: Why does he do this? The muck may be fetid mud, a pile of horse droppings or the putrefying carcass of a rabbit but the result is always the same.

Dogs, like donkeys and horses, also love to roll in the dust. The dominant sense of a dog is olfaction, the sense of smell. If dogs could speak they would doubtless express surprise that human beings fail to appreciate the delicious odours that they themselves find so attractive. The only answer to the question above is that dogs *like* rolling and if there is added to the exercise the experience of a penetrating odour their satisfaction is thereby heightened.

Jumping

The second comfort movement listed by the two Dutch scientists was called jumping. They described it in this way: 'The animal shoots [darts] very wildly and in an irregular course zigzags through the water. The

A male Grant's gazelle using its hindleg to scratch its head in a comfort movement. This is one of the characteristic gazelles of Kenya, found usually on semi-desert ground, where it commonly remains in the open during the heat of the day, yet is able to go without water for long periods.

movements are far too quick to be observed in detail.' Anyone who has kept aquarium fishes will know at once what they mean. In human beings we call it 'acting daft', which is one of those delightful phrases that literally mean nothing but which convey immediate comprehension to the listener.

'Acting daft' is most commonly seen in children. It is not encouraged and tends to die out as one grows older, yet every so often it is manifested in sober adults in their less inhibited moments. It is not *joie de vivre* yet seems to bear some relation to it. It is more a loosening of controls, physical and mental, in which stress is banished. We ritualize it in sport and in some of the more bizarre dances which are a feature of youth. Each generation – indeed, each culture – evolves its own dance form and enjoys it. All may, however, look equally ridiculous to the sober, objective observer. And all add to the enjoyment and the feeling of well-being of the performers. If nothing else they make the blood flow more briskly and this is the surest antidote to coronary afflictions!

Jumping can be exemplified by another, more or less inexplicable, feature of animal behaviour. It has been commonly observed that household cats, as well as the larger wild members of the cat family, behave as if they have taken leave of their senses in the presence of catmint. This plant is known as catnip in the English vernacular of North America; its Latin name is *Nepeta cattaria*; in French it is *herbe aux chats*; in Italian it is *herba dei gatti*; and in German *Katzenminze*. The mere universality of the association of cats with this plant, as indicated by these names, shows that people have long been aware of this intriguing aspect of cat behaviour.

The cat, large or small, wild or domestic, rubs itself luxuriously on the herb, rubs its face on it, and prances about as if it had taken leave of its senses. Essence of catnip has long been used in North America to attract the puma and the bobcat within gunshot or into traps. In the 1960s there were reports of a puma at large in Surrey, England. Fear of the Surrey puma reached fever heat, so the police were compelled to take action. Eventually it transpired that the 'pumas' that people were reporting were dogs and cats gone wild. One reason why doubt was cast on the presence of a real puma was that quantities of essence of catmint laid down where the 'puma' was reported to have been seen failed to lure any member of this large North American species into the open.

Meanwhile, two Canadian scientists were investigating the antics of cats in the presence of catmint. They found that the actions of a domestic cat in the presence of this herb were similar to some of those of a she-cat in heat. This is, however, inconclusive for the animals used in their experiments included young cats and old, males as well as females, normal as well as neutered or spayed cats. At the moment, the only

African elephant indulging in the luxury of a dust-bath. This action illustrates the multi-purpose nature of the elephant's trunk which while being an organ of smell, can also be used both for taking up water for drinking and dust for dust bathing, two very different functions which might seem incompatible.

suitable explanation is that cats like the smell of catmint, and it is known that essence of catmint has a soothing effect on the nervous system. So paradoxically a cat 'acting daft' after smelling catmint appears to be reacting to a feeling of well-being.

Tail-wagging

The third comfort movement listed from observing the fishes was called tail-wagging. All except the breast fins are closed against the body and the tail is beaten rapidly to and fro. The breast fins are moved as in backwards swimming, counteracting what would otherwise be a forward movement of the body from the driving action of the tail. So the fish remains stationary in the water while beating vigorously with the rear half of the body, that part known as the tail, or more precisely, the caudal peduncle. This is an extraordinary thing for a fish to do and we can only suppose that it is a true comfort movement.

Flapping and flicking

The next movements on the list have to do with flapping and flicking actions of the fins that seem to have their counterpart in the stretching movements seen in other animals. It hardly needs scientific investigation to convince us of the effect of these. In general we can be sure that one result is a quickening of the blood flow. Another is to relieve tensions in muscles caused by maintaining one body position without relaxing over a sustained period.

Anyone who has been digging the garden for a while, thus using a limited set of muscles, finds himself unconsciously stopping and

Electric blue damselfish, of the Central Pacific coral reefs, yawning and stretching. In fish these actions precede active movement. Damselfish are strongly territorial and will rush out and attack any other damselfish approaching their territories. This one has seen an intruder and is preparing for a quick dash to confront it.

straightening his back, and this action is refreshing.

The Bible says: 'Can a man by taking thought increase his stature?' The answer is that he can, if only by a fraction of an inch. Carried out in the correct manner the exertion is almost instantly invigorating. If carried out on a bed or on the floor one can almost feel the blood flowing more freely, accompanied by a feeling of well-being.

A stupendous yawn from a lion, an animal noted for its indolence. Yawning is normally a means of quickening the flow of blood in order to lessen sleepiness and fatigue. In lions, as in other beings that overeat, it has the added function of quickening the digestion.

Mumbling and nipping

Two of the last comfort movements on the list both have to do with movements of the mouth, other than yawning. One is called mumbling, in which the fish is said to make repeated biting movements with the mouth half-opened. The other is called nipping at the surface. In this, the fish, while swimming or floating near the surface, with its head directed upwards, makes nipping movements with the jaws. The movements are accompanied by a peculiar ticking sound, made by a sudden opening and closing of the mouth.

The Dutch scientists gave no more information about these comfort movements than are given here, so we are left to indulge in speculation. The two mouth movements seem to have some relation to preening and grooming which will shortly be considered in detail.

Yawning

Another item on the list of comfort movements in fishes is yawning. This is an action we normally associate with being tired and at first sight it seems to have little to do with comfort. We yawn when we are tired or bored and, in particular, we yawn when we see somebody else yawn.

Years ago a lady wrote to a newspaper about a picture she had hanging over her mantelpiece of a child yawning. She remarked that it gave her secret amusement to see how her visitors would start to yawn almost uncontrollably after sitting for a while facing the picture.

Since the Dutch scientists drew up their list more has been brought together concerning why fishes yawn. Surprisingly, they yawn most by day: when they are excited, or when they see a predator. They may yawn when they see food. Above all, fishes yawn when they have been inactive, either lying on the bottom or moving only slowly, for a long time and are about to move quickly.

It seems that yawning accelerates the flow of blood through the body. This tones up the muscles and makes available an instant supply of energy, so putting the fish into the condition when it is best able to move quickly, snap at food or escape from an enemy.

The biggest yawn in the world – unless the huge whalebone whales are given to yawning – is that of the hippopotamus. The only thing known about this is that it is the signal for attack. That is to say, it has been observed that when two hippopotamuses square up to each other for a fight they yawn hugely and then attack. What has not been investigated fully is whether in yawning a hippopotamus sets its blood coursing through its veins, so bringing its body into fighting trim, or whether the yawn is a ritualized signal of intention to attack, as in a baring of the teeth.

Wolves use ritualized yawning. There is a ceremony which normally occurs before they leave their lair after they have finished resting. It ensures that no wolf moves off before all are fully roused and ready to start. The domestic dog is believed to be descended from the wolf. Yawning is infectious with dogs. A human being yawning, or even pretending to yawn, in front of a dog will cause it to yawn in response. Cats do not respond to yawning, however, because it has no place in their social behaviour.

We yawn during the day if we are tired, and we yawn especially in the late evening and on rising in the morning. Yawning makes us fill our lungs, which drives the blood from the lungs to the heart. The increased supply of oxygen is taken to the muscles so reducing fatigue when we are tired. During sleep the rate of blood flow slows down and the blood itself becomes sluggish; the morning yawn sets the blood flowing normally.

This still leaves unexplained the yawning from boredom, known in humans and dogs, and possibly occurring in other animals. Fishes yawn if they are somehow thwarted; it is then known as displacement yawning. That is, when an animal hesitates between two courses of action its energy is displaced in a third that has no relevance to the real situation. A man listening to a bore, to whom he cannot be rude, is torn between the impulse to get up and go and the impulse to stand up and shout in protest. Instead he yawns and this is a further example of displacement yawning.

A baboon, similarly, will yawn when uncertain about what is going to happen. His yawn, however, has been ritualized into a show of teeth, or, in other words, a threat. At least, that is one suggestion put forward by those who study the behaviour of baboons. On the face of it, it seems odd that a comfort movement should become the reverse, a war-like act. It is like Darwin's suggestion that a smile, which we normally use as a symbol of goodwill and everything that is relaxing, is a modified snarl.

Preening, grooming and cleaning

Since 1950, when the term comfort movements first appeared, it has been little used except by ornithologists. One looks in vain in book after book on animal behaviour for any mention of it. Moreover, ornithologists have tended to apply it to such activities as stretching and yawning, rest and sleep. They also associate it with other activities of birds such as preening, bathing, sunbathing, dust-bathing and the general care of the body and maintenance of the plumage in clean and healthy condition.

They may be correct in doing so, although, as we have seen, comfort movements in the original sense included mainly, if not solely, movements that stimulate the blood flow. The others constitute what is more commonly called toilet activities, while rest and sleep were originally excluded. Perhaps the best way to approach this is to assume that comfort and toilet are so closely associated as to be inseparable, and that the supreme comfort activities are, indeed, rest and sleep.

One thing that must be kept well in mind is that there has been a continual evolution in behaviour. Whatever we do has its parallel especially in the things mammals do, and which, to a lesser extent, all other animals do as well. If fishes and birds show comfort movements we should expect to find them similarly in reptiles and amphibians.

ABOVE Two of the most precious possessions of most insects are the eyes and the antennae. Consequently, part of the daily routine of the bulldog ant is cleaning these two parts of foreign bodies, using the forefeet as combs, passing them over the large compound eyes or pulling the antennae through the clasped forefeet.

OPPOSITE Wood ibises enjoying a preening session. These birds are especially widespread over South America and have favourite resting places in which to preen themselves in the sun.

BELOW Most hummingbirds live in humid areas where there is water for both drinking and bathing. Both activities are essential to most birds, and bathing, seems also to be a pleasure.

Moreover, if vertebrates show similarities in behaviour, one with the other, there is no reason to suppose that these may not be found among invertebrates. Let us take a simple example and follow it through.

When a bird preens, it rearranges its feathers so that the plumage is in order. In the process it also oils the feathers, using the oil from the preen gland at the base of the tail. In the process it may shake itself, so getting rid of foreign particles including dust. Grooming is the mammalian equivalent of preening. The mammal shakes itself and uses its teeth to nibble the base of the hairs, suggestive of the mumbling and nipping at the surface seen in fishes. Instead of using oil from a preen gland it licks its fur, cleaning it with saliva. There is not a precise correlation between the procedures of bird and mammal, yet preening and grooming are roughly comparable. And it must surely be more comfortable to have the plumage neat or the fur lying properly.

Reptiles and amphibians need neither preening nor grooming because periodically they cast their skin; this is known as sloughing. All foreign bodies are thus disposed of and the new skin that has grown under the old, and which is exposed in sloughing, is in perfect condition. The

Wild animals, like the hippo (above), wild boar (opposite above), and hyena (opposite below) wallow regularly, and the advantages are several. To the cooling action of immersion in mud can be added other benefits. The layer of mud makes it harder for blood-sucking insects to reach their skin, and as the mud dries and flakes off it carries with it foreign bodies as well as loose hairs. In addition, there may be therapeutic qualities in the mud itself.

chafing of fishes seems directly comparable to sloughing especially where the animals rub themselves against solid objects to be rid of the old skin. Different animals must use different methods to the same end because their skins have different compositions.

It is commonplace to see insects cleaning themselves. They concentrate on the three most vital parts: their wings, eyes and antennae. Using their legs they brush these carefully to remove all foreign bodies that might be a hindrance.

If we now look at some of the lowliest in the animal kingdom we come to the sea-anemones. Very few people have studied these to see whether they use comfort movements or carry out any toilet procedure. There is, however, one study that was carried out carefully, on a sea-anemone living on the Great Barrier Reef of Australia. This anemone becomes covered with sand as the tide ebbs. Within hours all this sand is removed. Cilia, which are minute protoplasmic hairs on the skin, moving in unison, carry the sand grains from the upper surface of the anemone to the rim and drop them over the edge. The anemone certainly cleans itself, as surely as any bird does when preening or any cat does when washing itself.

Bathing

A baby bird after leaving the nest and beginning to feed itself will go to water to drink. It happens often that on first dipping its bill into the water it straightway starts bathing movements on land. It shakes its head, ruffles its feathers, flicks its wings and tail, as it will do later when immersed in

water and flicking water over itself. The fact that it does this while still on dry land, surrounded by dry air, merely because it has touched water with the bill, shows that the pattern of behaviour involved in bathing is innate, or inborn. It needs only the stimulus of the touch of water on the bill to set the bathing actions in motion.

All a bird's actions, whether of preening, scratching, stretching or yawning, are innate, as is almost everything else a bird does. So far as bathing in very cold weather is concerned, the evident enthusiasm for it suggests that something more than mere cleanliness is involved. It can only be supposed that the cold water penetrating to the skin gives a stimulus, perhaps a subsequent glow of warmth, that is highly satisfying. Only some such explanation can account for the way in which birds will queue for a place in a small puddle or jostle each other out to have the water to themselves.

Sunbathing is also innate to a bird. It is easy to imagine that the impact of the sun's rays on the body is health-giving, if only in imparting a feeling of well-being. What is more surprising is to watch closely a bird about to sunbathe. Often it has a favourite spot in which to perform. As the sun bathes this spot the bird will walk or run towards it, drooping its wings as it goes. Arriving on the spot it sinks to the ground, spreading its wings and tail and pressing them as close to the ground as possible. The most significant thing of all is the change in the eyes. They lose their lustre as if the bird is going into a trance and giving itself over completely to the joy of basking. Another manifestation of what appears to be a trance-like state is that the bird seems unwilling to end it. A bird that

European rabbit washing itself. During washing of the ears the anti-rachitic vitamin is transferred to the mouth. This is only one example of the few already known, and the many that will probably become known in the future, in which an animal's toilet benefits not only the outside of its body but its general health inside.

normally flies away when one approaches to within 10 ft (3 m) of it will not fly away when basking until one is 5 ft (1·5 m) or less from it.

Sunbathing seems almost universal among land animals. Lizards have chosen places where they indulge it, flattening their bodies to expose the maximum surface area of skin to the sun's rays. They also show the same reluctance as birds to run away, although the moment there is danger of their actually being grasped they come out of their trance immediately and are gone like the wind.

Grasshoppers make a great show of basking. They half-spread their wings, spread their legs and flatten their bodies to the ground. A grasshopper sunbathing looks so flat that it might have been crushed under somebody's heel.

Not a great deal of scientific attention has been paid to sunbathing but some years ago the matter was investigated in rabbits. Then it was revealed that a rabbit washing its ears was not merely responding to a desire for cleanliness. It was found that the natural oil of the skin, irradiated by the sun's rays, produced vitamin D which the rabbit takes into its mouth and swallows. If a rabbit's ears are kept clean by continual washing with ether, which dissolves the oil, the rabbit develops rickets; vitamin D is known to prevent this disease and it is believed that the irradiated preen oil of birds also has this property.

Sunbathing is not always obviously linked with the prevention of rickets. Nevertheless, there is the likelihood that in all cases a link with a health-giving process could be established if an investigation were made. Different effects can be achieved in different ways, according to the species or the type of animal implicated, without abrogating the principle involved.

Thus, while many birds show an almost ludicrous passion for bathing in water, some never bathe. Instead they use what is called a dust-bath. This is especially so with ground-nesting birds such as the domestic fowl, pheasants, guinea-fowl, bustards and sand-grouse. They disturb the dust with their feet, ruffle their feathers and contrive to kick the dust over themselves. This seems to clean the feathers of parasites, judging from the lice sometimes left behind on a dusting spot. It must also give the participants satisfaction, even pleasure or pleasurable sensations, since a bird settling down to dust-bathing shows the same trance-like look in the eyes and the same tolerance of an intruder as in sunbathing.

Dust-bathing birds range in size from house sparrows to the largest living bird, the ostrich. Among mammals that regularly dust-bathe there is a similar range in size, from mouse-sized gerbils to the horses and zebras. The largest land mammal, the elephant, enjoys the best of both worlds. It is given to bathing in water and also takes up dust in its trunk and sprays it over its body.

Gerbils roll in fine loose sand and groom themselves immediately after this to set their coats in order. Larger mammals are content with a shake. Birds follow a dust-bath with a combination of shaking and preening.

Animals and sleep

Sleep has been extensively studied in human beings since the 1950s, but it has been observed in animals for a lesser period and then mainly to

A characteristic of all members of the cat family is their ability to relax, especially when asleep, like this leopard sleeping off its meal in the crotch of a tree. Leopards prey on a wide range of birds and mammals. They also take fish and seem to be especially partial to a diet of dog. Like most carnivores they need to drink each day.

determine how long in each twenty-four hours a particular species sleeps.

Some animals seem never to sleep. They only doze, the eyelids being closed but only at intervals. Sheep rest with the legs tucked under the body and head erect. Goats adopt a similar posture although they seldom close their eyes.

Horses are said to sleep standing up but it is not uncommon to see one lying on its side with its head stretched on the ground, apparently in a deep sleep. In a herd of zebras only one member sleeps in this position at a time, and then only for a short while. Others may be dozing but always the bulk of the herd are watchful even if relaxed.

Domestic cattle have their rest periods when they crouch on the ground and chew the cud. During such a period they take it in turns a few at a time to drop their chins to the ground and close their eyes for a short nap. One of them will roll on to its side with legs stretched out and neck limp, and be fast asleep for five minutes. When it raises its head and draws its legs under it another will go into the position of deep sleep. Rarely does more than one member of a herd go into a deep sleep at a time. The same is true for antelopes.

The first systematic study of sleep in birds was made on ostriches in a zoo. They rested for seven to eight hours each night, squatting on the ground, with the neck erect and with eyes closed. They opened their eyes at the slightest sound. During this time one only would lie on the ground with legs stretched out and the neck limp and lying along the ground, and remain fast asleep for 5 to 10 minutes. When it ended its sleep and sat up, another would stretch itself along the ground and fall fast asleep for a similar period.

Already, from the few examples given, a pattern seems to be emerging: that animals rest and they also sleep; and sleep commonly takes two distinct forms, light sleep and deep sleep (we could talk of three forms if we include the sleep of hibernation).

There is another pattern. If we compare the small amounts of sleep habitually taken by elephants, horses and cattle and the amount taken by predators in the wild and by domestic animals and compare these with the sleep of monkeys, apes and man, we arrive at a general conclusion. It is that the more highly organized the brain the more is the amount of sleep taken in every twenty-four hours. Experiments in the last few years have confirmed this, and they also indicate that the better the brain the more is dreaming needed. Dreaming is now seen to be a process in which the brain, our computer, sorts itself out, in preparation for another spell of activity and of recording experiences.

That even birds may dream is suggested by the way in which pet talking parrots, when apparently fast asleep will mumble the words they have learned. Just as in our dreams events are jumbled, so the sleeping parrot will jumble the words from phrases used while awake.

Going down the animal scale, it is certain that fishes sleep. Some even hibernate. And parrot-fishes not only sleep but their bodies prepare for it by giving out a slime that forms an envelope around the body – a sort of nightdress that is presumed to be protective in some way.

Amphibians and reptiles certainly rest. Whether they sleep has yet to be determined. Going lower in the scale, that insects rest is a matter of everyday observation.

So, starting at the bottom of the scale, we see that the lowest animals rest. Sea-anemones withdraw their tentacles and contract the body periodically, so alternating periods of activity with periods of quiescence (that is, rest). Any invertebrate we watch for any length of time shows this same alternation of activity and quiescence.

It could be, although this is a shot in the dark, that true sleep only occurs when the animal possesses a true brain. That is, sleep is a characteristic of vertebrates. In that event, we can suppose as a consequence that rest is a means of refreshing the body whilst sleep is necessary to refresh the brain.

The great sleepers

Anything said so far about sleep consists of generalizations, and as with all generalizations there will be many exceptions. First, there are the exceptions due to individual variations; for example, some people require much more sleep than others.

Another exception is the hibernators. The hazel dormouse, of Europe, can be most agile and alert when occasion demands. Yet even by day it seems to fall into a doze the moment there is no acute need for movement. Added to this, it spends half the year in hibernation, in a sleep

The hazel dormouse, the proverbial sleeper, spends little of its time fully awake. It hibernates for half the year and spends three-quarters of the other half either asleep or dozing. Even when active it is always ready to drop into a doze the moment the need for active movement ends.

near to death, its body cold to the touch, hardly breathing and with its heartbeat all but stopped.

In Alaska, in the frozen north of America, lives the Barrow ground squirrel, a sort of chipmunk. The Barrow ground squirrel sleeps underground for nine months of the year. During the brief summer, beginning as the snow starts to melt, it wakes. It cleans out its burrow, looks for food, mates and raises a family, gorges itself to lay in fat for the next winter, expands its burrow and takes in new bedding. Although it works feverishly during this short season of activity, as though every moment were precious, it nevertheless takes a regular few hours in each twenty-four to rest.

The arch-sleeper of the animal world is perhaps the bat, and especially the small insect-eating bats of temperate regions. Bats hibernate for six months in each year. They sleep by day and come out to hunt by night, which means, in the summer months, that well over half of each day/night period is spent resting and sleeping. In addition, no bat spends the whole of the night on the wing. There are spells of an hour or so flying, hunting for insects, followed by periods of rest of two to three hours before taking wing again, and so on through the night.

Nobody has yet determined the ratio of rest and sleep to flying activity in each twenty-four hours for any insect-eating bat but a fair guess would be that relatively little time is given to active work. Taking the year as a whole, so including the six months' hibernation, a bat probably spends upwards of 80% of its time asleep or in a state of drowsiness.

6
The Social Graces

The moment an individual enters the world, as a larva, hatchling or baby, it has to adjust to its environment. The more simple the animal, in structure and behaviour, the less is the adjustment needed. It profits us most, therefore, to look mainly at those in the higher reaches of the animal kingdom because that is where the adjustment will be most marked and, incidentally, where it will be nearest to that which we and our children have had to make, so providing a basis for easy understanding. Nowhere is the picture complete, however, because in spite of the mass of zoological information already recorded there are still many gaps in our knowledge.

If, then, we start with the mammals we are dealing with an individual, a newborn baby, that in a sense awakens from a long period of sleep within the womb where everything was provided for its comfort. It was kept warm and fed regularly, kept clean and cushioned against accident, injury or the harshness of the environment. Then, suddenly, it is ejected into a new world: it is born. The most immediate effect felt is the lowered temperature. It loses its protective coat, the placenta. Its food supply is cut off suddenly, and it must start to move about on its own.

Survival would be impossible if such a baby did not carry with it an inborn pattern of behaviour and an ability to improve on this by experience. But learning by experience takes time. In the meanwhile the mother, largely through a pattern of behaviour released by the process of motherhood, contributes vastly to the early progress of her offspring.

A normal mammalian mother with a litter of 'helpless' young licks her babies as a matter of routine. As the babies develop they catch the habit, often licking the mother as she is licking them. As they grow older, they lick each other. When a parent licks a baby, or vice versa, they are strengthening the bond of affection between them. This is also the case when the growing babies lick each other. This makes for peace between them, which is particularly necessary in multiple litters. Yet, as they grow older still, other emotions begin to stir. They, like children in a human family, begin to quarrel, which is the process known as smoothing the corners off each other. If it is objected that comparisons between animal and human babies are not justified one can point to the many ways in which bear cubs are very like human babies. One outstanding trait is that when feeding or resting they are apt to develop mannerisms or tricks of behaviour which they insist in carrying out. Then suddenly they will drop them, just as human babies will. One bear cub, for example, that was being bottle-fed, insisted for a short period on swinging one hindleg as it stood up to take its bottle and refusing to feed if not allowed to do this.

Teaching the young

This smoothing off of corners is especially important in a social species, such as the wolf and the African hunting dog. It is part of the education of the young, and it raises the question of how far animal parents teach their young the ways of life. In all teaching, whether of human or animal young, the best lessons are those of setting an example and of encouraging the young to do the correct thing. In the main, animal parents do no more than that, and even that amount of education only

OPPOSITE Male jaguar showing affection to its mate by licking. Jaguars look very like leopards, their counterpart in Africa and southern Asia. The most obvious distinguishing feature is that a jaguar typically has a spot at the centre of the rosettes in its coat.

figures at all prominently in the young of carnivores. There is one bird which more than most birds seems to control and marshal its young especially in times of danger and in the search for food. This is the common starling.

Once young foxes begin to take solid food the parents, especially the father, go out hunting. He has the tendency, when he has returned with food in his mouth, not to give it to the cubs immediately. The result is that they have to jump up to snatch the food as the father holds his head high. His actions may be due to a waning of the unselfishness that seized him when the cubs were first born, so that his actions now can be seen not so much as deliberate teaching but as an unwillingness to relinquish the food he holds.

Later, when the cubs are old enough to follow the parents on hunting expeditions, they learn by example. This is probably true of all hunting mammals. The young learn not only what prey to hunt but what tactics to use in getting near enough to seize it. They will build on this knowledge later, from experience. That imitation of the parents seems likely follows from what is known of the red fox.

One reason why this handsome carnivore – so useful to man in his agriculture and forestry because its staple food is destructive mice, voles and rats – is persecuted is that it has a reputation for killing poultry. It seems, however, from the writings of knowledgeable people, that poultry-stealing is localized. There have been foxes known to live on or near poultry farms that have never killed a bird. What happens is that if a

OPPOSITE LEFT Female guanaco, one of the South American camels, in an affectionate encounter with its lamb.

OPPOSITE RIGHT The coyote or prairie wolf shatters the silence with its social howling.

BELOW Brown bear family with mother playing with one of the cubs.

vixen feeds her cubs on poultry they in turn become poultry-killers.

There is on record an observation of a polecat which had its kittens in captivity. When the kittens were a month old and their eyes had opened, the mother would take a piece of meat, with which she was fed, place it in the nest, drag it out again and leave it lying on the ground. The kittens then followed the scent and found the food.

Similar behaviour has been observed in the meerkat, a species of mongoose in South Africa. In this, when the mother finds something to eat she will run up and down with it in her mouth, when the young ones have just come out of the nest and are beginning to take solid food, and they have to snatch it from her. If they fail to take it she will lay it down in front of them. One consequence of this is that they learn what foods to take. A litter born in captivity were offered banana and refused it, until the mother picked it up, when they immediately ran up and snatched it from her and soon became extremely fond of the fruit. Their normal food is mainly insects which the mother catches, runs to her youngsters and lets them snatch from her mouth. In this way she introduces them to a wide variety of insects.

Keeping the young in check

There are, however, scattered records of occasions on which a wild animal parent has seemed to be 'teaching a lesson' to its young, and of doing so by chastising it. It is often said that a bitch will teach her puppies to snap. Whether it is in fact an educational matter or merely play that resembles snapping, it does sometimes happen that an exuberant pup will bite the mother harder than it should. She replies with a nip of her teeth or a blow from a paw which elicits a yelp of surprise rather than pain from the pup, suggesting that it may have learnt its lesson.

The best example of animal education of the young is found in the hippopotamus. According to a detailed account given by an experienced Belgian zoologist, the male hippopotamus is very apt to attack any young hippo whereupon the mother defends it. If the attack is made in water, the most secure position for the young hippo to be in is swimming beside the mother and level with her shoulders. It seems that she insists on her infant always accompanying her in that position.

On land, any encounter with a male is likely to be face to face. The most secure position for the young one then is following the mother. Certainly, young hippos on land normally follow the mother and a female has been seen, when her youngster strayed from this position, to beat it unmercifully with her huge muzzle until it squealed. She continued this chastisement until it obediently followed behind her.

Female elephants have been reported to tear off a branch of a tree or uproot a sapling with the trunk and belabour a recalcitrant calf. No scientific study has been made of this to determine the circumstances so as to judge whether this behaviour is educational or merely occasioned by an irritable mood in the mother. The usual interpretation is that it is some form of punishment used to teach the young good manners.

Such examples belong to the category known in human dealings as corporal punishment, that is, the systematic and deliberate infliction of pain intended to be used as a deterrent. The corrective methods used by animal parents towards their young are far less spectacular and much less easy to describe. They resemble more nearly the methods used in the experimental laboratory for testing the behaviour of animals and which are known as the method of reward and punishment. That is, the animal being tested is encouraged to perform a certain action and is rewarded with food if it succeeds and is mildly punished, either with a tap on its body or a light electric shock if it does not succeed.

Animal parents, and these remarks are mainly confined to mammals although they do apply to some extent to the actions of some birds, correct their youngsters by giving them mild punishment and following it almost immediately with a soothing gesture. For example, dog and cat parents will punish a misdemeanour by giving the infant a quick and gentle nip on the neck or face, or a slight pat with the paw. In cats, the pat can be varied by using the paw with the claws withdrawn or with the claws slightly out. In dogs, the parent may give a puppy a kick that sends it somersaulting across the ground.

Such punishment evokes squeaks in the infant but no cry of pain. Moreover, it is almost invariably followed by the parent licking the offender, as if to indicate that there is no malice and only an educational value in the punishment.

It is possibly a warped outlook that makes the human observer look for corporal punishment in any form in the natural upbringing of animals. Here it has been suggested that mild punishment is used but perhaps it should be stressed that the showing of affection is at least equally important and probably significantly more important.

ABOVE Tasmanian devils are friendly and sociable. Here two of them meet almost nose to nose in a greeting. The nose to nose greeting is the equivalent, in many animals, of the greeting kiss.

Codes of behaviour

On the outbreak of war in Europe, in 1939, it was anticipated that there would be massive air-raids on towns in Britain. So a plan was set in motion to evacuate all children to rural areas and put them into the care of foster-parents. Two such children, a brother and sister, from one of the poorer districts of London, were taken in by a middle-class family. These children were dirty and dishevelled, and they lacked all forms of social graces.

Children so evacuated were subjected to a medical check-up soon after their arrival. The local doctor described them as being like wild animals. Evidently, they were from a household in which parental love and devotion were at a premium. The doctor told how he had given each a penny – almost a fortune to a child in those days – as a peace offering. Each snatched the coin and retreated promptly to the chairs on which they had been seated. When the doctor next approached they made something like snarling noises and each sat on its penny. A proper medical examination was impossible.

Three months later the doctor visited the children again. He told how it was hard to believe that they were the same boy and girl he had seen on his first visit. Loving care on the part of the foster-parents and the inculcation of good manners had transformed them.

There are several implications to be seen in this simple episode. In human societies each family has its code of behaviour which we call their manners, and these may be labelled 'good' or 'bad' according to whether

OPPOSITE Brazilian tapirs keeping cool and indulging in mutual grooming, nipping each other with their teeth in the manner of members of the horse family. Tapirs are shy, solitary and seldom seen. They are virtually defenceless and when disturbed make for the nearest water, swimming underwater if necessary to escape.

they create harmony or disruption in daily life. The code, as a result of social intercourse, becomes general among members of a community enjoying a similar environment. It becomes noticeable between different sections of a community. It also varies from one part of a nation to another, as it does between nations and, on a broader scale, between one ethnic group and another.

Social and non-social groups

The human element has been introduced to establish a base from which the ways of animals can be examined to see how far anything comparable to the social graces of people can be detected.

The first principle to emerge, in any wide-ranging survey of the animal kingdom, is that in the great majority of animals behaviour is innate, that is, it is inborn or instinctive. Instinct can be modified by learning. The ability to learn increases as the nervous system, particularly the brain, becomes more advanced. It follows also, quite logically, that the need for good manners, or social graces, are least necessary in solitary species and at their maximum in social species.

Before going further, it should be emphasized that not all animals living in groups are social. Swarms of male gnats dance in groups where the females join them for mating. Flies congregate on a rock warmed by the sun. Moths gather around a light. At best these can be said to be gregarious. People in a crowd watching a football match, for example, are also gregarious. They have come together temporarily and by chance. A society on the other hand is a group of animals which remain together for long periods and which react to each other's behaviour in the best interests of all concerned.

A social group of rhesus monkeys in India, their harmony broken by no more than a yawn. The wide-open mouth, exposing the teeth, would in any other context suggest a threat display, the prelude to a fight.

There are social insects, such as bees, wasps, ants and termites. Many species of vertebrates are social and live in shoals, flocks, herds and schools. Insect societies are based on a division of labour, with a fertile female (queen) and one or more castes of sterile workers.

The peck-order

Among vertebrates there is invariably some kind of dominance hierarchy, usually spoken of as a peck-order. In this hierarchy, relationships between the members of the group are established and maintained. One benefit of this is that it lessens the chances of surprise meetings with other members of the same species, which might startle or enrage them, so psychological stress is reduced to a minimum.

The dominance hierarchy was first discovered and studied, in 1913, in domestic hens, when it was noticed that in any group there was one hen (A) that could peck any one of the others without retaliation. Then it was seen that a second hen (B) could peck all the others except A; C could peck all except A and B, and so on down the line to the one that could be pecked by all the other hens and would not peck back.

The peck-order is established by fighting and an inferior can rise in the social scale by winning a fight with a superior. A stranger, introduced into a flock of hens, will be attacked by all the others and has to go to the end of the line unless it can secure a place higher up by fighting for it.

This has long been known to poultry-keepers who introduce a strange hen into a flock by putting it in the roost at night. Farmers who keep cows have known, long before 1913, that cows in a herd have a similar

A macaque grooms the fur of her baby. While cleaning and grooming are primarily hygienic functions, they can assume the importance of a social ritual when younger or lower-ranking members of the species groom their superiors and meet no resistance.

hierarchy. This is most clearly seen when the cows are being brought in from pasture to the stable, for milking. One cow, the boss cow, always takes the lead, and there is one cow that always brings up the rear. The rest have their recognized places in the file and any cow that gets out of place is butted by the others until the accepted order of precedence has been restored.

Once the news of the discovery of the peck-order had been impressed on the scientific world, and publicized in the Press, zoologists began to look for dominance hierarchies in other species. Today, we can say with confidence that they are found in most vertebrates, never in social insects, although there is a suspicion that there is something like a dominance hierarchy in bumble-bees.

There is also a dominance hierarchy in multiple litters. Where there is more than one young in a litter, one of them is usually found to be dominant and the rest sort themselves into order of precedence.

It is possible there may be a link here with another observed fact that in any multiple litter not all will be equally tameable. Even when taken to be hand-fed before the eyes open, some members of a litter will be less manageable than others. Presumably, left in the natural state, these would have been dominant over their more amenable litter-mates.

A hierarchy such as that seen in domestic hens is known as a linear hierarchy. There are others types. In baboons, three inferior males can dominate one boss male. Although he may be able to defeat any one of the three he is unable to remain dominant so long as the other three support each other.

In baboons, also, adult males are dominant over sub-adult (half-grown) males, sub-adult males are dominant over females, and juveniles are subordinate to all males and all females. Among gorillas, the younger males accept the older males as having a higher status and do so without fighting. A fully adult male gorilla has silver-grey hairs on its loins. Younger gorillas recognize this as an indication of status and do not challenge him. A mane and larger canine teeth mark a male baboon as fully adult and usually the sight of these is sufficient for younger baboons to accept him as superior.

A peck-order has been seen in some freshwater fishes, mainly those kept in aquaria where their behaviour can be readily observed. Similarly, a peck-order has been demonstrated in some lizards, and it may well be proven in the future that these hierarchies are universal throughout the vertebrate kingdom.

Fishes certainly learn to recognize each other by sight. In an aquarium a superior-ranking fish will spread its fins at a subordinate swimming near. This is a threatening gesture which serves as a warning to the subordinate to behave with due respect.

There are two other ways in which aquarium fishes indicate their rank in the social order. In the first, the lowest ranking individual occupies a space near the bottom of the aquarium, the boss is near the top. The fishes pass through the spaces occupied by those lower in rank, so that only the boss has a space to himself.

In the second, rank is indicated in some species by the posture of a fish when two meet. The lower ranking is shown by the fish tilting itself obliquely upwards. The more it tilts, with the head directed upwards, the lower its rank.

There is a peck-order in jackdaws, another social species of birds. Lorenz, the distinguished German student of animal behaviour, has shown that jackdaws pair for life and that if a low-ranking female pairs with a high-ranking male she takes on his rank. He also found that a high-ranking female will not pair with a low-ranking male. A low-ranking female having risen above her station through marriage must, however, establish her position by fighting those females previously superior to her, and she does this in competing with them over food.

ABOVE Prairie dogs at the mouth of the tunnel leading into their coterie alert to the approach of a stranger. It has long been said that prairie dogs post sentries to warn the rest of the colony of danger. But it seems that any prairie dog sitting up in moments of relaxation and sighting danger bolts for safety. Before doing so it whistles, thus warning its fellows to make themselves scarce.

'Women and children first'

It may be that in many species battling for food is the most common way of sorting out a position in the hierarchy. In domestic hens it has been found that when grain is thrown to them it is the boss hen that pecks at it first, followed by the others in order of precedence.

Other things than food are also involved. In a troop of baboons a female in heat is taken over by the boss male, who endeavours to keep all other males away. Once the heat has waned he loses interest, she loses her status and any male may attempt to mate with her. A female with an infant is also treated with respect. She is not attacked, which is the equivalent of being at the top of the hierarchy. There have been several stories of a male baboon attacking a female until he sees she has a baby, which up to then had been hidden from his view. He immediately drops all aggression towards her.

OPPOSITE Wolf cubs playing at the entrance to the burrow in which they were born, watched by their mother. Play is necessary not only to develop the senses and the muscles but also for strengthening the social bonds between members of a family.

The social order is clearly shown when a baboon troop is on the move between a feeding ground and a sleeping ground. The old males are at the centre of the troop together with females on heat and those carrying infants. On the periphery of the troop are the youngest, and therefore the weakest males. These give the alarm when a predator is sighted. The adult males come forward to show fight, leaving the females and infants within a protective screen formed by the young males.

Rhesus monkeys have a similar arrangement. A female with an infant is treated with respect, and so is an aged male. The dominant male does not permit other males near the females gathered around him, but in troops of crab-eater and bonnet monkeys – nearly related to the rhesus monkeys, in which the proportions of the sexes are more nearly equal – subordinate males are allowed near the females provided they show proper respect. They do this by adopting a submissive attitude and smacking their lips, which says in effect: 'My intentions are honourable'.

One feature in the behaviour of baboons and monkeys that stands out is the consideration shown for the female. Generally speaking, it is true of most mammals, as indeed for humans, that males show a chivalry towards the females. In some species it is so marked that even if the female attacks a male he will refrain from doing her physical harm.

Harmony and disharmony

In the story of the wartime evacuees earlier in this chapter, reference was made to manners differing with the environment in human society. In some species of monkeys just mentioned, the pattern of behaviour

Herring gulls squabbling over food. This well-known seabird lives and roosts in flocks, and one gull finding food soon attracts the others by its movements or its calls. This seeming harmony can be readily broken when two of them reach out for the same morsel of food.

within the troop also changes with changes in the environment but, as in human beings, mainly in small and unimportant details.

All accounts of social organizations studied in monkeys, including baboons, emphasize that sexual relationships are closely linked in the dominance hierarchy. Yet dominance can occur at an age preceding sexual maturity. This may be partly because some individuals are more aggressive than others as shown by the differing degrees of tameability in multiple litters. So dominance or otherwise could begin in the cradle.

There is, however, the contradiction that the well-being of young animals is more dependent, in a social sense, on brothers and sisters, however much they may quarrel, than on the parents. At least, that much is suggested by the somewhat repellent experiments carried out in the laboratory with young monkeys.

These experiments were mainly of two kinds. In one set, a baby monkey was offered an artificial mother made of wire mesh with a mask representing a face. Bottles of milk with rubber teats were placed where the mother's nipples would be. A guaranteed supply of food was, however, not enough for the baby. When it was offered a pair of wire-netting mothers, complete with feeding bottles, but with one bare and the other covered with warm towelling, the baby always chose the latter. While feeding it would cuddle into the towelling which represented for it the mother's fur.

In the second set of experiments baby monkeys were kept isolated from their playmates. It is hardly necessary to describe in detail the distress they showed. It suffices to say that they showed the pitiful symptoms we should expect. Moreover, young monkeys so deprived showed marked anti-social characteristics as they grew up, even delinquent tendencies.

The moral of our story about the two wartime evacuees now becomes apparent. Mother-love – even father-love as a substitute, as a second best – can not only determine an infant's happiness but also its physical and mental well-being, as well as its ability to live in harmony in a society. With maternal love must be linked a degree of training, but the love is more important to the infant's overall well-being.

Given a normal upbringing, why should young animals quarrel? The answer is that they are battling over food. To illustrate how great a part food plays in the lives of animals we can turn to a series of observations on Barbary sheep.

The experimenter used twelve sheep in a large enclosure in the Bronx Zoo in New York. The sheep were, therefore, not in close captivity and, although domesticated, their basic natural instincts were likely to have full play. The twelve were made up of four males, four females and four juveniles. All were marked with coloured spots for easy identification.

When two sheep were close to the fence, food was put down half-way between them. At sight of the food both moved towards it, but the more dominant of the two always got it by threatening the other or by actually butting it. The tests were continued for nearly three months. At the end of that time it was clear that males were dominant to females and both were dominant to the juveniles. Moreover, the males showed an established hierarchy: A took food from B, B from C and C from D. It was the same with the females and also with the juveniles.

This kind of experiment merely puts on a statistical basis what naturalists are led to expect from watching wild animal families or what farmers are convinced happens in the case of their domestic animals.

People who keep animals as pets, allowing them to breed, and watch the progress of the growing young tell the same story. They add something to it, however. This is that they get the impression that sometimes the parent will intervene to check the greed of a dominant youngster so that all the family get their fair share.

With some carnivores the open mouth is used as a threat but it can also figure in play. These Himalayan bears are expressing pleasure. This double meaning of the open mouth is especially noticeable in another carnivore, the fox, which opens its mouth in courtship.

Teaching manners

The question of whether young animals learn or are taught manners, in the human sense, has not been explored scientifically. It may be that this can only be done by random observation rather than direct experiment.

The kind of random observation can be illustrated by reference to a family of blackbirds. In this species the babies are altricial, therefore fed at the nest until they are fledged. Then they leave the nest. The parents continue to feed them for about two weeks. After this the family group breaks up because the youngsters have learned to feed themselves.

Occasionally, one sees a young blackbird soliciting food of the parent after the time when this should have ceased. The parent finally loses patience with the youngster continuing to beg or even snatching food from her beak, and she attacks her youngster, driving it away. Unless we are misinterpreting what we see, the young blackbird has been taught to go away and find its own food.

In human societies children are guided mainly by vocal instructions, facial expressions and gestures. Animal parents, especially among the higher mammals, use much the same means and these are the result of an inborn pattern of behaviour. Special study has been made of the facial expressions of wolves and of the messages conveyed by the position and posture of the tail. These, together with the vocalizations, represent a means by which one member of the species gauges the mood or intentions of another and constitute a language. The youngster's response to such signals is either innate or readily learned by experience.

7

Territory and Courtship

There was a time when poets and lyricists wrote enviously of the freedom conferred on birds by having their forelimbs modified to wings. They were, of course, not referring to ostriches, penguins and kiwis which are earthbound but to the common or garden birds that, free as the air, roamed wherever their fancy took them. At the same time naturalists were noting the opposite.

As bird-watching became a scientific study, naturalists began to be aware that if they traversed the same route day after day, at the same time of day, they were very likely to see the same birds. The idea began to develop that birds, like every other living being, were creatures of habit, tied as if by invisible threads to the same patch of ground.

The establishment of territory

It was left to the English ornithologist Eliot Howard to put this dawning discovery on to a firm basis. He noted that with the coming of spring the winter flocks of reed warblers began to break up. Each male sought out an area of reeds where he perched and sang. At intervals he would return to the flock but as the season wore on he spent more and more time on his own little patch, until finally he no longer returned to the flock.

As the male reed warbler grew more and more attached to his territory, as Howard called it, so he sang more frequently and with greater volume. He also became less tolerant of his fellow reed warblers. In fact, he became distinctly aggressive towards any of his fellow males that approached his singing post. It became evident to Howard, as a result of his careful and patient researches, that each territory had boundaries as fixed as those of a fenced property.

If another male reed warbler approached this boundary there would be a skirmish between him and the occupant of that territory. Should he cross the boundary there would be a fight. With birds there is more fighting over boundaries than over any other single cause.

Boundary fights between the neighbouring warblers established the confines of the territories; and into each territory came a female, drawn there by the song of the occupying male. If she liked the look of the territory and of its owner she stayed and the two became a pair. The female also learned the boundaries of the territory and she took part in defending them, especially against an invading female.

Within the territory the pair built a nest, mated, the female laid her eggs and together they fed and reared their young. There was, however, one jarring note. When the female first entered the territory the male attacked her, as he would have attacked any trespassing male. The attack would be short-lived, for whereas an intruding male showed fight when attacked, a female adopted an attitude of submission. The belligerence of the occupying male died down and he recognized her for what she was, a possible future mate.

Some birds do not mark out their territories until after the nest is built, but all react the more strongly to intruders of their own species once the nest-building has started.

One striking fact has been confirmed by numerous observations, namely, that territorial fights do not go in favour of the stronger. In human affairs, possession is nine tenths of the law, and we all feel a greater

OPPOSITE The red deer stag is a mountain of force and fury in the period preceding the rutting season. Then his roar heard at close quarters can be quite terrifying. It is a signal that the stag is gathering together a harem of hinds and warning other subordinate stags not to interfere.

This tiger snarls at the photographer not because he regards him as prey but as an intruder into his territory. However, each year a number of people are killed by tigers. Fishermen working in the rivers of India seem to be specially vulnerable to tigers when coming in to the bank.

sense of security and have greater courage once we are inside our own property, whether that property be a single room or a large estate.

Any occupying bird confronted by a stronger, more powerful intruding neighbour, may at first be forced to retreat. But as he is driven nearer to his nest his courage mounts accordingly while that of the intruder tends to wane. There is an unwritten law, shared by birds and men alike, as well as by other animals, that another's property is sacrosanct and that his boundaries must be respected. The law may be broken at times but normally it prevails; it is embodied in what has now become a familiar term, the territorial instinct.

Since 1920, when Eliot Howard first drew attention to his far-reaching discovery, more research has been directed towards the

territorial instinct, and the aggression arising from it, than to any other subject in biology. From the many studies made, it is now established that most vertebrates and many insects defend territories.

Another equally important principle has also been established. It is that generally speaking males do not, as the romantics have led us to believe, fight over the females. They fight over territory into which they will take, lure or invite the females. The illusion has arisen because all too often when two males are fighting over a territory there is almost certain to be a female nearby. In many instances the reverse is true, that the females fight – but not necessarily with the same outward display of vigour – over the male.

Male golden pheasants of Asia become very aggressive at the beginning of the breeding season. The fights are territorial and only incidentally over the possession of the females. The theory is that the gorgeous plumage attracts the female and its display stimulates her to breeding condition.

Sizes of territories

Although the general pattern of the acquisition and holding of a territory, described by Howard for reed warblers, is typical there is much variation as from one species to another. Thus, for small song-birds it may vary in size from as little as 1200 to 2400 sq yd (1000 to 2000 sq m) to 2½ acres (1 hectare).

The territory may be used only for mating or nesting, as in the gannets which nest in densely packed colonies with only enough room between the nests for the gannets to walk to reach their own. Each nest is defended aggressively by its owner; in fact neighbouring nests are sited just out of pecking distance by the sitting birds.

Gannets do not need more territory since they feed out at sea, whereas birds with nests spaced apart from each other use the territory as a source of food, wholly so in some species, partially so in others.

In some species, sparrows and finches are examples, there are no territories beyond the edges of the nest. Birds of prey have vast territories over which they make sorties from the central nest in search of food. Herring gulls neither mate nor feed within their territory which, nevertheless, is usually a large area around the nest. In some colonial nesters, all the females band together to defend the territory occupied by the colony.

Some territories are used for feeding alone and these are defended outside the breeding season. The European robin is a well-known example. During the breeding season a pair occupy one territory. In winter male and female occupy separate territories for feeding only.

Gould's manakin uses a territory for mating only. The male clears a 'court', an area of 20 by 30 in (50 by 75 cm), of every leaf or twig, leaving only the bare earth. He stays near it throughout the day, driving away any other male that shows signs of approaching it. Should a female wander near it he begins an elaborate display to attract her attention. If she is beguiled, she enters the court and the two begin a complex dancing display, both leaping into the air, after which they fly away together to make and build a nest.

Although the idea of territory was first identified in birds and is most strongly associated with them, certain insects are known to defend territories, as are some fishes and reptiles as well as many mammals.

The evolution of territorial instinct

It is of interest to pause for a moment, when considering any basic form of behaviour, to examine it from an evolutionary point of view. In this connection it cannot be emphasized too strongly that it is important to keep in mind the gross structure of the animal kingdom, from the lowliest single-celled animals to the mammals, as represented by the system of classification (see opposite).

BELOW Red swordtails, tropical freshwater fishes, have a marked peck-order. In this picture two males are sparring, swimming backwards towards each other, their bodies arched, displaying their tails to settle their ranking in the hierarchy. The swords are no more than elongated anal fins and are therefore only decorative. Although the females lack a sword they still display at each other in the same way as the males.

OPPOSITE BELOW Female laughing dove of Africa is taking a bath, and as she does so the male continues to court her, displaying from the rock above with his breast pulled out in characteristic pigeon fashion.

With this in mind we can contemplate, for example, how asexual reproduction, which figures so strongly in the lives of the lower animals, is gradually lost as we ascend the scale. Equally, the related phenomenon, the regeneration of lost parts, is very strong in the lower animals and peters out at the level of reptiles, in lizards, which cast their tail as an escape measure and then grow a new tail.

The territorial instinct, on the other hand, shows a contrary pattern. It is most marked in the higher animals such as the vertebrates and is largely absent in the lower invertebrates. This is mainly because most of them live in water where their food is brought to them by currents or it swims to them. Crowding is possible because food is plentiful.

In the case of animals living on land food has to be sought, as a rule, so living space becomes necessary and there must be some mechanism for ensuring an adequate food supply to each individual. This is especially necessary when young have to be fed. Consequently, parental care which begins low in the animal scale and increases markedly with birds and mammals is linked with the intensification of the territorial instinct.

Among invertebrates a territorial instinct is most marked in dragonflies each of which can be seen patrolling its beat over water, driving away intruders by flying at them. It is also a noticeable feature of some butterflies, the males of which settle on particular leaves and fly out to intercept other males flying through their air-space. This may, however, be more a matter of sexual activity, each male keeping a space clear for himself into which females may wander.

Fishes that show parental care, either by building a nest, as in sticklebacks, or by mouth-brooding, also have a strong territorial sense. Most of these are freshwater species. As in birds it is the males mainly that guard the boundaries, which are so well defined that an occupying male, swimming out from the centre, will stop abruptly at the boundary as if he had banged his snout against an invisible wall.

Classification of the Animal Kingdom showing the main subdivisions with approximate number of living species

VERTEBRATES
Mammalia (mammals) 3700
Aves (birds) 9000
Reptilia (reptiles) 6000
Amphibia (frogs, toads, newts) 3000
Osteichthyes, Chondrichthyes, Agnatha (lampreys) } (fishes) 21,000

PRE-VERTEBRATES
Cephalochordata (lancelets) 30
Urochordata (sea-squirts, salps) 1600
Hemichordata (acorn worms) 91

INVERTEBRATES
Pogonophora (no common name) 100
Chaetognatha (arrow-worms) 50
Crustacea (crabs, lobsters, prawns, etc.) 30,000
Insecta (insects) 1,000,000
Chilopoda (centipedes) 2800
Diplopoda (millipedes) 8000
Pauropoda (no common name) 400
Symphyla (no common name) 120
Pycnogonida (sea-spiders) 600
Arachnida (spiders, ticks, mites) 60,000
Merostomata (king crabs or horseshoe crabs) 4
Onychophora (peripatus) 100
Echinodermata (starfishes, brittlestars, sea-cucumbers, sea-lilies, sea-urchins) 44,000
Annelida (ringed worms) 9000
Echiuroidea (echiuroid worms) 150
Sipunculoidea (peanut worms) 250
Mollusca (molluscs) 128,000
Brachiopoda (lampshells) 260
Phoronida (horseshoe worms) 15
Bryozoa (moss-animals) 4000
Entoprocta (no common name) 60
Acanthocephala (thorny-headed worms) 650
Aschelminthes (rotifers and various small worms, mainly parasitic) 12,500
Nemertina (ribbonworms) 750
Platyhelminthes (flatworms) 13,000
Ctenophora (combjellies) 80
Cnidaria (formerly Coelenterata) (sea-anemones, sea-firs, jellyfishes, corals, sea-pens) 10,000
Parazoa (sponges) 2500
Mesozoa (degenerate flatworms) 50
Protozoa (single-celled animals) 50,000

ABOVE Anole lizard displaying aggressively at his own reflection in a mirror on the assumption that his mirror image is another male intruding into his territory. The lizard will then go about his daily activities but will return from time to time to display at his reflection.

RIGHT Male fiddler crabs have one claw enormously larger than the other which they wave to indicate possession of a territory and to attract a mate. When the tide goes out, exposing a tropical sandy beach, the crabs come out of their burrows to feed, and at intervals each male flashes its large claw. The large claw is about 1½ in (3.7 cm) long, almost twice the width of the body.

Red fox vixen laying a scent trail while licking the dew from her muzzle. The dog-fox will pick up her trail to find her. The scent glands are in anal sacs and on the upper surface of the tail, near its root, but the urine is probably the main vehicle for scent secretions.

The home range

The pattern is different in mammals. Many defend a small area vigorously, which is the territory proper, and use a larger area around it as the home range in which they feed, but this home range may be shared with others of their own species.

As with birds of prey, the beasts of prey need a much larger territory than herbivores, especially when these are grass-eaters rather than eaters of any soft-stemmed plant. In one national park it was ascertained that a lion's territory measured 1–1½ sq miles (2·5–4 sq km) whereas even large herbivores like the hippopotamus were satisfied with a territory one third this size. In this same park, by contrast ⅓ sq mile (1 sq km) would support 24 kob antelopes, 12 topi antelopes and 2 reedbucks.

Even small mammals can hold a large territory where, as with the woodmouse, it is held by a group. A dominant woodmouse male marks out a territory of 4–6 acres (1·5–2·5 ha) and this is used by the whole colony. The boundary is regularly patrolled.

Howler monkeys of South America live in clans which commonly consist of around three males, four females, four juveniles and three infants. They range through the trees over a given area and any other clan intruding into this area is greeted with a low barking roar that reverberates through the forest. Howlers call in chorus at dawn and near dusk, a habit which, it is believed, tells all howler monkeys around which territories are occupied.

On the ground, many mammals, especially the hoofed animals follow a well-worn network of paths through the home range. These paths lead to recognized pastures, water-holes, wallows and excreting places. Only the territory proper is defended and the grass is not grazed on it. This has the advantage that the long grass provides cover while the animals are resting and off their guard.

This tendency not to feed on the territory, as distinct from the home range, has led to an interesting situation. A fox leaving its earth in the evening will play with a leveret or pass a pheasant without molesting it

Spectacular nuptial displays of the male sage grouse attract the females and bring them into breeding condition. The males fly to the display grounds at 4 a.m. There each erects its tail fan of twenty spiky feathers and inflates the two orange air-sacs on the breast, the display being accompanied by mincing steps.

provided that they are within the territory. On the home range both would be treated as food. Anteaters that feed on termites will make a burrow in a termite nest located on their territory but will not molest the insect occupants.

It seems almost a rule that no animal hunts until it is a certain distance from home, and the rule seems to hold good for birds of prey since small birds often build their nests under those of hawks and eagles. This appears to give them protection from other predators, and at the same time they are immune to attack from the bird of prey itself. Even human hunters tend to go well afield from their base before starting a hunt.

Control of populations

The holding of territories also helps to control the population numbers. The best documented instance is of grouse in which breeding success depends on the size of the territories. Each autumn the males battle for a share of the available heather-covered land. A number of territories are marked out as a consequence and the losers, up to half the total numbers of the males, are driven out to the periphery. There they fall prey to predators or die in some other way. By contrast, very few of the pairs that occupy the territories fall victim to either predators or natural causes. They therefore survive to breed.

Probably the most specialized territory-holding mammal is the prairie dog, a species of North American ground squirrel, which lives underground but comes up during the day to feed on herbage. At their height they were spread widely over the prairies but their numbers have declined in the face of human settlement. Some idea of their former numbers can be gained from an estimate of one prairie-dog 'town', made in 1901, that was said to cover 100 miles (161 km) by 240 miles (386 km) with a population of 400 million. A town is made up of separate 'coteries' (groups) each of which possesses its own territory that is communally held against members of neighbouring coteries. Each coterie comprises a dominant male, perhaps a subordinate male, three females and a half a dozen juveniles. The coterie system of the prairie dogs is a clear example of territorial behaviour keeping a population spread out and evenly distributed. Through the antagonism between members of neighbouring coteries living space of at least 3–7 ft (1–2 m) is maintained between each where the herbage can be grazed by these burrowing rodents.

Because the members of a coterie are related and, more particularly, because they are living together, they carry the same odour. When two prairie dogs meet, they 'kiss'; that is, they meet head on and touch noses. The action is no doubt indicative of friendly intent but it probably also helps in ensuring that both have the correct smell. Any stranger, one that does not smell right, is driven out.

Whereas birds, being sight-animals, recognize and maintain the boundaries of their territories by reference to landmarks, most mammals mark their boundaries with an odour. They urinate at special points or deposit their droppings in selected places, or else use secretions from special glands on their bodies for depositing a perfume. Antelopes and deer carry such glands on the face or between the hoofs. In carnivores, such as dogs and cats, these glands are known as the anal glands and are situated below the root of the tail. These special glands may, however, be situated on almost any part of the body, according to the species. Some species have such powerful scent glands that their odour is all-pervading, and the animals are named accordingly, for example, the musk-deer, musk-ox and musk-rat.

Courtship in the lower animals

As we have seen, one of the basic uses of a territory is to ensure a supply of food. Probably more important is the part it plays in courtship.

It is not sufficient for the survival of a species that there should be so many males and so many females each contributing their sperms and ova. For successful breeding the sperms and ova must be made available simultaneously. In lower animals living in the sea this is brought about by a spawning crisis.

As the breeding season approaches the ovaries in the females and the testes in the males ripen. This means the ova and the sperms are maturing. When all is ready, a signal is supplied by one starfish, sea-urchin, marine worm or whatever the species in question. The animal sheds its ova or sperms into the sea and with them a chemical messenger, a pheromone, that spreads through the water and stimulates all the other members of the species to spawn.

Courtship fulfils this same purpose. It is a means of bringing male and female into breeding condition so that they are ready to mate.

The fruit-fly is a tiny insect much used in the laboratory for research into genetics. It has no territorial behaviour. At mating time the female selects her mate, a male who has been searching for a female of his own species by touching any small insects he meets with his forelegs, seeking the chemical stimulus which only a female of his own kind can give out. Once they have met he places himself close to her, usually standing behind her but with his head towards her. Then he circles her, keeping close to her and all the time facing her.

During these manoeuvres the male extends one wing sideways and moves it vigorously up and down, producing a stream of air that strikes the female, arousing her. From time to time he licks her then tries to mate with her. If she is not ready she kicks him away. So it continues until she is fully in breeding condition and permits the union. Courtship and mating may take only two to three minutes or it may take much longer.

Female moths in breeding condition give out a perfume. They vibrate their wings rapidly, causing a stream of air that carries the perfume into the surrounding air where wind currents bear it farther. Molecules of the

Each earthworm is male and female so mating is a mutual exchange of ova and sperms, and courtship is unnecessary. Provided two worms are in easy reach of each other, in neighbouring burrows, they need only to stretch out across the intervening space for the union to be consummated.

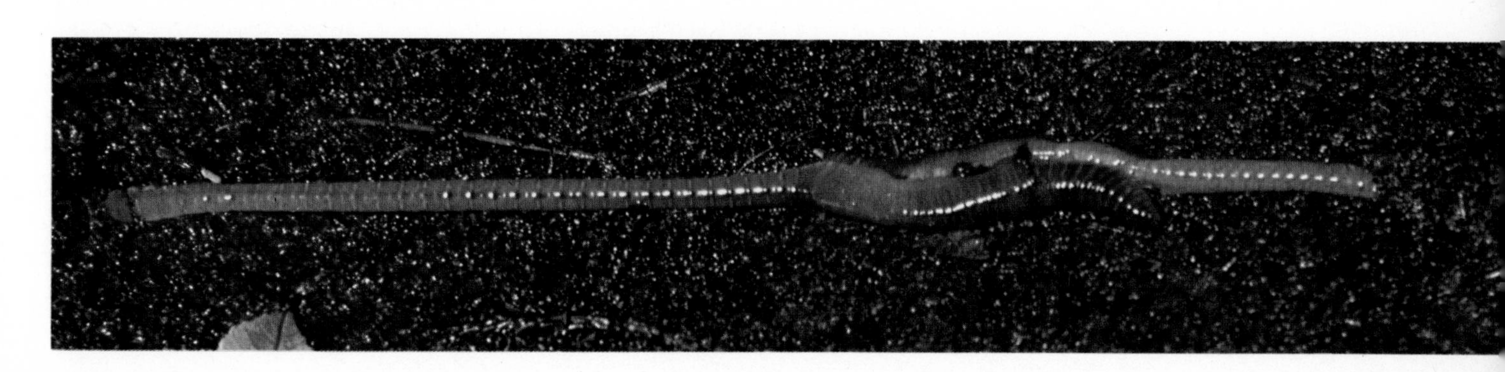

perfume falling on the antennae of the male, perhaps a mile (1·5 km) away, draw him to the spot where the female is waiting.

The male grasshopper sings. The female hears him and is stimulated. She turns her head in his direction and walks to him, and then they mate.

Web-spinning spiders rely on vibrations for their food. A fly caught in the web struggles and vibrates the silk. The spider feels the vibrations in the threads, comes out from her resting place and seizes her prey. The male looking for a mate twangs one of the threads of the female's web with his front legs, thus attracting her attention. In many species of spiders there are special spines and files on the legs that produce a particular tune on the web of the female.

Hunting spiders, that make no web, use visual signals to woo the female. The males have special decorations which they display in a sort of dance. Their front legs are gaily coloured, and they raise them and wave them. They may jerk the body or perform a dance. The patterns of courtship are almost endless, varying from one species to another. In all of them, the male appeals to the sense of sight.

In one species of spider, the male, lacking colour or other personal adornment, catches a fly and wraps it in silk. This he presents to the female and while she is unwrapping it he mates with her. Certain small flies that feed on even smaller flies have a similar courtship. One species goes further and presents his mate with no more than a silk wrapping. While she is unwrapping the bogus parcel he mates with her. These small flies, and also female spiders, make courtship dangerous for their males, being apt to eat their suitors since they are so voracious at all times, hence the stratagems employed to placate them or distract their attention.

The male stickleback builds a nest of soft water-plants, and by spring he has grown a red patch on his breast. A female seeing his red 'throat' is stimulated to tilt her body so that he sees her belly swollen with eggs. He swims towards her, leads her in a zigzag dance to the nest and, jerking his head and body towards the entrance, shows her where to go. She enters, leaving her tail still exposed. He butts it with his snout. She spawns,

The satin bowerbird builds a bower of sticks and decorates it with coloured objects and in it displays to his hen.

Wandering albatrosses spend the year at sea, going to oceanic islands to breed. Their courtship displays include showing off their magnificent wings and fencing with their huge bills. In most of the smaller relatives of the albatrosses, including the petrels and shearwaters, the courtship displays take place at night and consist of noisy flying displays. That of the wandering albatross takes place in daytime and is accompanied by dances, pecks and chirps.

leaving the nest through the other side. He enters and spawns in his turn, thereby fertilizing the eggs.

Successful mating, that is, successful fertilization of the eggs, usually depends upon a delicate synchronization of the physiology of the reproductive systems of the two sexes. Courtship displays serve to bring the two into equal levels of readiness. By the use of colours, sounds, perfumes, dances, gifts and the like, the partners to the matings are excited and the hormones, together with innate behaviour, ensure a successful outcome. So delicate is the balance that female birds have been known to lay on seeing the reflection of their own displays in a mirror.

All aspects of reproductive behaviour become more complex as we go up the animal scale. The nervous system becomes more highly organized and the sense organs more finely discriminating. There are also more indications among the higher mammals of associated affectional behaviour and, at times, something looking uncommonly like what, in ourselves, we call love and falling in love.

The exhibitionists

Courtship has been most thoroughly studied in birds and for good reasons. Sight is the most important sense in birds, as it is in humans, with hearing second. With few exceptions the sense of smell is very poorly developed. Consequently, as we should expect, colour and sound predominate in most bird courtship. Yet having said that one is reminded that one of the most successful species of bird, and one which achieves a high breeding success, is the house sparrow, with unusually drab plumage and unmusical voice. Nor does it have any spectacular

displays, no dances, graceful pirouetting, courtly bowing or strutting that are such strong features of bird courtship as a whole.

Indeed, in no other group of animals are such extravagances seen as in bird courtship. In the great majority of species the signs that a courtship is being enacted are no more than that two birds are behaving in an unusual way. One, the hen, is the less active. The other, the cock, is more sprightly than usual and tends to circle the hen, spreading and quivering his wings, flicking his tail, bobbing and bowing and appearing almost to dance.

For each member of a species the courtship display is the same, its pattern stereotyped or specific to the species. In addition to the bodily movements and the posturings, the male makes the most of any colour in his plumage. The colour may be no more than a yellow bar in each wing which normally is inconspicuous. In the display the bar is exposed as the wings are spread and not only made more conspicuous but the feather or feathers comprising the bar seem to be larger.

A striking example of how birds under emotional stress during courtship alter their appearance is seen in the common magpie. Normally, it is pied, with about as much black as white showing in the plumage. In display the white areas are increased so that instead of a pied bird we see what looks like a white bird with a few small patches of black. The transformation is most noticeable.

The same is true of a crow, a bird that normally appears an uninspiring black. In courtship display the feathers of wings and tail are spread and, seen close to, even this dull bird temporarily appears magnificent. The male bird also shows every sign of excitement, which is doubtless communicated to the female he is courting. Indeed, we can go further and say that the excitement is caught by every member of the species that is within sight of it, so that there is a kind of 'display crisis' comparable to the spawning crisis of sea-urchins.

We can also be sure that, just as we are moved to rivet our attention on a male bird courting and to describe its appearance in superlatives, the female he is courting cannot remain unmoved. Her blood, we can suppose, courses wildly through her body, her muscles tingle, her hormones are working overtime, and her gathering excitement mounts to the point where she responds by an invitation to be mated. She is, in more prosaic terms, stimulated into breeding condition.

There is evidence that in colonially nesting birds this general excitement is vital to successful breeding, and when the numbers are reduced, so that the excitement is at a low ebb, they fail to breed altogether, or at best they breed with less than normal success.

There are some species, notably birds like the capercaillie, that have an unusual pattern of courtship. At breeding time the males gather at a special arena known as a lek, stamping ground, court or other such name. They display special feathers, inflate coloured wattles or sacs, and give every appearance of intense excitement. They are quarrelsome, on edge, and occasionally one will limp as if afflicted by a temporary paralysis of the legs. The females enter one at a time to select a mate who follows her to a short distance from the lek where they mate.

In species in which male and female look alike, so that it takes an expert to tell one from the other, both display on equal terms,

OPPOSITE Many birds successfully attract a mate although their plumage is drab and their song most unmusical. At the other extreme is the peacock with its extravagantly beautiful train that can be raised and spread in an enormous fan.

Rose pastel canaries, the male feeding the female as part of courtship, an action often seen in wild birds and which is continued in domestication. The action is symbolic but it probably tells the female whether the male is a desirable mate, one who will tend her with devotion while she is sitting on her eggs.

ABOVE The courtship display of the male Jackson's whydah is one of the most remarkable sights of the African savannah. The sparrow-sized male leaps and prances to attract the female.

The brilliant colours on the male mandrill's face intensify as its excitement grows, whether that excitement is the result of aggression or love-making. The colours are less conspicuous in the young males, becoming more pronounced with age.

performing the same or similar actions. Both contribute to the mounting excitement. This may express itself in a wild dance, as in the cranes, with the birds prancing on their long legs, spreading their feathers and leaping into the air. The variations are endless.

At the other extreme are the peafowl, the birds-of-paradise and the many others with flamboyant and richly decorated plumages. It is highly doubtful whether the females of these species experience any greater thrill, or nervous excitement, on being courted than the hen house sparrow with her drab mate. Nor is there any reason to believe that the breeding success of the species is the greater for all the display of colour, art-forms and general beauty of the males. At any rate the excitement engendered by these outrageously elaborate plumes is no less than that created by more modest displays.

So far as the male bird-of-paradise is concerned, he puts everything into his courtship. He whistles, he prances about on the branches, he fluffs out his gorgeous plumes to the full and, in some species, ends by hanging upside-down, adding grotesqueness to beauty.

In all this display, the female, who is drably and inconspicuously feathered, behaves as if wholly unimpressed. The coyness of the female in the presence of the male's extravagant exhibitionism, until really aroused, has long puzzled the naturalists.

In the bower-birds, of Australia and New Guinea, the female has a sober plumage and is quite unremarkable. The male is little better. What the male lacks in gay plumes, however, he makes up for with his bowers.

A typical bower consists of a platform of sticks into which is implanted a double row of vertical sticks to form an avenue. This is decorated with snail shells, small bones, coloured berries, flowers or leaves. The walls of the avenue may be painted with the coloured juices of ripe berries.

Other bower-birds carry their bower-building to even more extreme lengths. They build what can be described as an open-fronted house with a garden in front bounded by a hedge of sticks. The garden is finally decorated with blossoms snatched from local plants.

It is as if the bower-bird, lacking extravagant and highly decorated plumage, has used materials lying around to put on a show to impress his future mate. Within his bower the male carries out his displays. Having done so he mates with her, and she departs and is not readily seen again. She is too busy building unaided a nest as much out of sight in the shrubs as she can make it.

There are elaborate courtships in some fishes and reptiles, especially in those that have colourful bodies, but in striking contrast the great majority of mammals do not have anything like these spectacular displays, and for a good reason. Most mammals are smell-animals in which smell is the dominant sense. In them colour vision is either lacking or poorly developed and sight is only a secondary sense.

Courtship in mammals

It has been shown that badgers indulge in much play prior to coition and at the same time there is a strong smell of musk. Foxes also play during the pre-mating period. Elephants show strong signs of affection, usually

expressed in stroking each other with their trunks. Some members of the weasel family and some mongooses indulge in boisterous, semi-serious fighting prior to mating, especially when the female is not fully receptive. If she repulses the male he may grasp her firmly by the neck, sometimes drawing blood, but always he licks the neck copiously where he has seized it. These and other observations point to a courtship in mammals which may be prolonged or may be very brief, but which is less ostentatious than in the sight-animals such as birds.

The exception among mammals is the primates, which include lemurs, monkeys, apes and man. In these, sight is the dominant sense and it is significant that colour and body ornamentation are conspicuous features of many monkeys. Even in monkeys with drab coats, colour is a marked feature of the genital regions and the buttocks. The extreme is reached in the mandrill with the brilliant red and blue stripes on the face with similar colours on the genitalia.

Perhaps the most surprising colour pattern in mammals is that of the gelada baboon, a monkey closely related to the mandrill and drill. The breast and nipples of the female are bright red and so shaped that that area of the body resembles closely in colour and configuration her genital region – which is seldom seen because she spends so much time sitting!

Young sable antelope bull in the preliminary to mating which is known as displaying *flehmen*, a German word used to describe a characteristic facial expression in the bull. Having sniffed the cow he draws back his upper lip, exposing the teeth, as if in an expression of disgust. *Flehmen* is used in most hoofed animals, as well as in some others, such as cats. In it, it is believed, the male is determining whether the female is receptive to mating.

8

Wheels within Wheels

At two hours before midnight on 11 October 1492, as the frail ships under the command of Christopher Columbus were approaching the American continent, the men on watch saw a mysterious light out at sea.

At the Marine Biological Laboratory at Plymouth, England, in the 1930s a scientist was watching sea-anemones and was struck by the idea that throughout their lives these animals are performing a slow-motion underwater ballet.

In the early 1950s just before eleven o'clock on any Tuesday morning a dozen or so household cats could be seen gathering together at a street corner in the heart of London.

These three apparently wholly disconnected events are, in fact, closely related, and each has to do with rhythmic behaviour and the biological clock. These two phenomena have been familiar to mankind since the dawn of time but only since the beginning of this century has attention been focused on them.

Rhythmic behaviour

Rhythmic behaviour is the term applied to actions repeated at regular intervals which may be as long as a year, as in migration or breeding, or a matter of seconds as in the beating of the heart. Some of these actions seem to be controlled by external circumstances and are referred to as exogenous. Others can only be controlled from within the body and are called endogenous.

The most familiar examples of rhythmic behaviour are those that are repeated daily, within the span of each period of twenty-four hours. We are generally aware that most birds start to move about at dawn, are active throughout the daylight hours and go to roost as night begins to fall. We say they are diurnal, that is, active in the daytime. We hear owls hooting at night and we see bats come out just before nightfall. They are said to be nocturnal, or active by night.

When we pursue the matter further we are astonished to note the unfailing regularity with which bats leave their roosts in places where light penetrates so little that we can justifiably say that the bats have been in total darkness, and should be unaware that the daylight is fading. But something within the body of the bat has told it that night is at hand.

The cats gathering at eleven o'clock every Tuesday on the street corner were reacting to another inward mechanism, a time sense. At that time and on that one day of the week there arrived somebody, known in those days as Pussy's butcher, who threw scraps of meat to the cats. So on Tuesday, and on Tuesday only, the cats gathered punctually to receive their expected treat.

A cat or a dog fed punctually each day at, say, five o'clock in the afternoon will make its way precisely at that time to its food bowl.

One of the best examples of an animal showing a time sense was the dog belonging to the famous botanist, Linnaeus. The dog always went to church with its master. Whether Linnaeus aided and abetted it or whether the dog initiated the action will never be known, but if the sermon went on too long the dog would either bark or leave the church half an hour after the sermon started. When Linnaeus was ill the dog still went to church – and still left when the sermon had lasted half an hour.

OPPOSITE Serried ranks of gannets on an isolated island responding to the annual cycle of hormone activity which urges them to mate and nest. The nests are made of seaweed and are close together, often touching each other. Only a single egg is laid by each female and both partners of the pair incubate it.

Tidal rhythms

The behaviour of animals living on the seashore is even more fascinating. Each day they are uncovered by the tide twice a day. To avoid being dried up, shellfish must close their shells, and sea-anemones must withdraw their tentacles and contract their bodies. It looks as if they do this as soon as the water recedes leaving them exposed. If, however, these same shore animals are removed and placed in an aquarium in which the water level is always the same they will continue the same tidal rhythm. As those on the shore close with the ebbing tide so will those placed in the aquarium close despite no change in the water level.

The time of the tides, however, alters with each succeeding day. But the shore animals in the aquarium show the remarkable ability to alter their time correspondingly. So clearly it is not entirely the movement of the sea-water that controls their actions but something inside themselves.

Fiddler crabs live on the shores of tropical seas. Some of them change colour as the tide ebbs and flows; they become darker as the water uncovers them. The change in colour is brought about by pigment cells in their skin contracting and expanding. If a piece of this skin is removed and kept alive in salt water, under a microscope the pigment cells can be seen to contract or expand in time with the tides.

Tides themselves vary in size. At the new and the full moon the sea comes farther up the beach at high tide and goes farther down the beach at low tide. These are called the spring tides. Some oysters release their larvae into the sea ten days after each spring tide during the breeding season. This means their eggs are fertilized at the time of the spring tides.

Cockroaches and moth larvae living on the bat guano on the floor of the Batu Caves in Malaya. They are in perpetual darkness yet maintain the bodily rhythm seen in other species of cockroaches living under normal conditions.

Correcting for sunset

Cockroaches begin to be active just before nightfall. Their activity rises to a peak in the early part of the night and then drops to a low level half-way through the night and remains at that level throughout the following day. This rhythm can be upset by changing the external conditions, by changing the hours during which they are illuminated and kept in darkness.

Thus, experimentally, cockroaches were kept under a bright light for twelve hours. The light was switched off at 6 p.m., leaving them in total darkness. Once the insects had become accustomed to this, the light was switched off at 7 p.m. After a day or so the cockroaches adjusted to the new time. Using this method it is possible to have cockroaches in the same room that show different rhythms of activity.

Without this ability to adjust their times of activity, cockroaches would find themselves coming out too early in summer, when night falls at a later time. So what we have here is an internal rhythm which is influenced by external conditions. To put it another way, the internal rhythm, or biological clock, is corrected by the external factors.

Circadian rhythm

It is only some thirty years ago that it was first realized that small mammals, such as mice and shrews, are neither wholly diurnal (active by day) nor wholly nocturnal (active by night). Their lives are made up of alternating periods of three hours: throughout the twenty-four hours of

Eurasian harvest mouse feeding on green wheat grains. It obeys a three-hourly rhythm throughout night and day, active for three hours and sleeping for three hours. The periods of activity during the night are slightly longer than by day and the periods of rest slightly shorter. This is especially so during the summer.

each day they continually spend approximately three hours being active and three hours sleeping.

So much information has now been gathered about these rhythms that only a few examples can be selected to illustrate the different types of rhythms that have been discovered. The three-hourly rhythms of mice and shrews, for example, are known as a circadian rhythm. This is from the Latin word *circa*, meaning about, and *dies* meaning a day. No rhythm is of precisely twenty-four hours' duration – the tidal rhythm is an example – but about twenty-four hours.

Internal rhythms

Now we can turn to the slow-motion ballet of the sea-anemones at Plymouth. The species investigated was the plumose anemone. This is white or pink with the usual column-like body ending above in a feathery crown of tentacles, the whole about 3½ in (8 cm) high.

In a sea-water aquarium containing several dozen of these anemones, it is noticeable that at any given moment all will be in different states. Some will be still, others waving their tentacles. Some will have contracted to little more than buttons, others will be stretched to full height. There will be a few bloated like balloons, others long and slender.

When a single plumose anemone is placed on its own in an aquarium it can be seen that it is in a continuous state of very slow movements. When watched against a black background and its outline drawn at frequent intervals during the day and night the extent of its movements can be

OPPOSITE A long-eared bat typical of the small insect-eating bats that come out at night to hunt for insects. Even when these bats roost in almost total darkness they nevertheless come out punctually at nightfall guided by their internal rhythm. The light varies in intensity according to whether the evening sky is overcast or bright. The internal rhythm of the bat adjusts for this, at the same time making allowance for the day-to-day changes in the time of sunset.

recorded: body lengthening or shortening, tentacles shortened and thick, long and extended, still or moving, will all be observed. All movements are gradual so one gets the impression of the flowing grace of a dancer.

These movements do not necessarily have anything to do with the daily movements needed for feeding or escape from danger. By keeping the anemones in conditions of constant temperature, in water free of any food and undisturbed by vibrations, it was demonstrated that this rhythm was inherent; it was like continuous clockwork which, like a heart-beat, started on its own and could continue independently of what was going on around. When it was filmed and the film was shown speeded up, the full grace of the 'dance' could be appreciated.

The lugworm is a marine bristle worm that spends its life in a U-shaped burrow in the sandy shore. At the surface the position of the burrow is marked by a shallow depression and, a short distance from it, a heap of worm-castings. The lugworm normally lies at the base of the burrow, and continuously takes in through the mouth sand which lies at the bottom of the shaft going down vertically from the saucer-shaped depression. Hence the periodic sinking of the sand in the saucer. The sand passes through the body and is periodically ejected through the opening at the head of the other vertical shaft of the burrow. Periodically also, water is driven through the burrow, in a direction from the tail to the head, by the action of muscles rippling along the body.

This worm has a regular rhythm of forty minutes. Even when enclosed in a U-shaped glass tube in a tank containing filtered water, so that there is no food, no sand and no disturbances to distract it, the lugworm faithfully goes through the motions of taking in sand, aerating its gills and emptying its intestine. And it does so every forty minutes because it is responding to an inherent rhythm.

We see something of this rhythm in ourselves when we sleep. Theoretically, while asleep for the night we ought to lie throughout in the same position as when sleep overtakes us. Instead, even in the most peaceful sleepers, there are small bodily movements due to our inherent rhythm of activity, which has been compared to the engine of an automobile ticking over until the driver lets in the clutch. In the plumose anemone letting in the clutch would be represented by food landing on its tentacles or a predator touching it.

Lunar rhythms

The oyster releasing its larvae after the spring tides is connected with the first of the three events with which this chapter started. To solve the mystery of the light seen by Columbus's men we must look to a certain Caribbean worm, which swarms to the surface to spawn. At the surface the females attract the males to them to fertilize their ova, by lighting up and flashing the lights which are part of their bodies. The surface of the sea, over a wide area, becomes luminescent or, as it is commonly but wrongly called, phosphorescent.

The palolo worm, which lives in crevices in corals in the seas around Fiji and Samoa, is very well known to zoologists. As the time for breeding draws near the worms, male and female, produce their ova and

Resplendent weaver-bird of Africa, relative of the drab house sparrow, in nuptial dress and well on with the task of building its elaborate cup-shaped nest with a long entry tunnel. The nest is built by interlacing pieces of grass and vegetable fibre and is suspended from a tree branch. The long entry tunnel probably protects the young birds from predators, particularly snakes.

sperms respectively. These are located in the hind end of the body. On successive nights, these hindparts break away, wriggle out of the crevices and rise to the surface where they burst. The ova and sperms commingle and the ova are fertilized. The local people know exactly when this is going to happen. In fact, it occurs for one or two days before dawn at the third quarter of the October and November moon. They go out in their boats ready to scoop up the mass of spawn, which they eat.

Now we can begin to see why the breeding cycles of all animals are so regular, and why bird migrations work to a fairly rigid time-table. It used to be said, in the days prior to the massive extension in the study of animal behaviour that has been a feature of the twentieth century, that birds knew from instinct when to migrate. To some extent that is still true provided we substitute 'internal rhythm' for 'instinct'. Undoubtedly, as it is easy to observe, birds begin to be restless as the time for migration approaches. Experiment has also shown that during this time they tend to orientate themselves in the direction in which they will ultimately travel. The actual departure is, however, triggered by the length of the day.

When we speak of a 'fairly' rigid time-table it is because the pattern does not follow precise dates of the calendar. Thus, observation by countless ornithologists, and the records they have kept, tell us that there is a good deal of 'scatter' in the time-table. The common swift that migrates between Africa and Europe is one of the best avian timekeepers. The main stream of swifts arrives in Britain each spring in the last few days of April. The earliest arrivals have appeared, however, on 21 March. Similarly, swifts depart in late August but there are records of departures as late as 21 December.

Biological clocks

There must be remarkably few man-made clocks that keep absolutely accurate time. Some lose a little each day, others gain, even when somebody is constantly on watch to correct them. The internal rhythms (or clocks) of animals are equally unreliable. The length of day acts to synchronize them but, like the human clock-keeper, cannot always control the eccentricities of the mechanism. On the whole, the clocks of lower animals, such as the palolo worms, are more reliable than those of the higher organisms because their works are simpler.

In temperate regions some animals hibernate. This is supposedly to guard against a shortage of food in winter. Snakes, lizards, frogs, toads and insects go into a winter torpidity (which is not a true hibernation) and they will enter their winter quarters at regular times in autumn. The same is true for all other hibernators, and all are apt to do so even when the weather remains mild. The hibernators do not suddenly feel drowsy and, therefore, move into winter quarters. Preparations will have been going on in their bodies for weeks beforehand, in anticipation of the day when a drop in temperature triggers off the hibernation.

Breeding cycles

Other annual cycles are seen in the breeding season. Temperature triggers the breeding in some species, in others it is the seasonal rains. Whatever meteorological factors are involved, to precipitate the breeding, the fact remains that they only give the final touch, the trigger that brings to a climax preparations that have been going on inside the body for some time previously.

In birds, many mammals and some lower vertebrates, the reproductive organs regress at the end of the breeding season. That is, the ovaries in the females and the testes in the males grow smaller and become functionless. Just before the next breeding season the process is reversed. The reproductive organs increase in size to their former proportions and breeding is again possible. All this is part of an annual rhythm which is controlled and governed by hormones.

Except for those animals, the great majority, as we have seen, that shed their eggs and take no further interest in them, the preparations continue. Females that bear altricial (helpless) young especially need to make provision for them in advance of the birth. They need to make a nest. Even hyenas and foxes, which use a burrow and sleep on bare earth, depart from this normal behaviour as they feel their babies stirring within them. The female hyena, whose burrow is normally one taken

ABOVE LEFT The jabiru, a South American stork, responding to the urge to build a nest, is flying with a stick to lay in its foundations. The nest is usually in a tree, is large and made of strong branches intertwined with smaller sticks, the whole lined with grass. It is built 60–80 ft (18.2–24.3 m) above the ground.

ABOVE RIGHT Female turtle laying her eggs in the sand on a tropical beach. She may have travelled a thousand miles (1609 km) or more to reach the beach. She is urged to start the journey, and largely directed all the way, by the complex workings of a hormonal cycle.

over from an aardvark, takes grass in and arranges it as a cosy bed. The vixen, like the rabbit she preys upon, strips the fur from her front to receive her cubs.

The action of stripping the fur is a good example of how the internal workings of the mother keep pace with the development of the foetus. We can trace the course of events from the moment of conception to understand this better. From that moment when the sperm enters the ovum to produce a fertilized ovum the mother's body and the workings of her body begin a series of events that work smoothly to a time-table. The timekeepers are the chemical messengers or hormones which enter the blood-stream and are carried around the body.

One part of this time-table includes bringing the mammary glands to readiness for the production of milk. On receipt of the appropriate signal from the appropriate hormone they begin to enlarge. So do the teats, or nipples, which house the ducts through which the milk will flow.

We can surmise that these revolutionary changes in certain structures of her body do not pass unnoticed by the female. There is probably some form of irritation that makes the mother tear at the hair that surrounds the nipples. The result is that she accumulates a mass of hair and at the same time lays bare the nipples making it so much easier for her babies, when born, to find the source of their nourishment for the succeeding days, weeks or months, as the case may be. She makes the mass of hair into a nest, being stimulated to do so by hormonal actions which change and direct her behaviour.

As soon as the first discovery of these chemical messengers was made, which was in about 1906, a key was provided to a fuller understanding of events in the animal world that had been seen but which puzzled those who saw them. For instance, it was known before this that when a female hippopotamus was about to be delivered of a calf, she did something she had never done before.

Hippopotamuses live in rivers. They come on land to feed and rest on sandbanks. In no sense do they furnish their sleeping quarters. Yet for several hours before the birth of a baby hippo the mother spends her time gathering reeds and trampling them to make a cradle. This does not always happen, however. There was one occasion, for example, when a female hippo was seen to have her baby in shallow water at the river's edge. On another occasion a female gave birth on land, about a hundred yards from the water's edge, on bare earth. When the men who watched the birth went to take a closer look at the baby it got to its feet and ran to the river. Usually, however, the mother makes a reed cradle. The instinct to make a cradle must be there, as suggested by the pregnant hippo in a zoo, near her term. She was seen to gather straw that was in her pen, carry mouthfuls of it to her bathing pool to wet it, then carry it into the pen and trample it ready for her baby's birth.

All these things were noted and recorded before the full extent of the workings of hormones was known. Although they have not been investigated in the hippopotamus we can reasonably conclude that it is not the sight of the reeds alone that touches off the impulse to make a cradle, otherwise the female in the zoo would not have acted as she did.

It might be argued that there was something peculiar about the hippo that dropped her calf so far from the water's edge. There is evidence that some mammals can delay the moment of birth if immediate circumstances are not favourable. Statistically, the best evidence for this comes from domestic cats. Many cat owners will tell how their cats will not deliver their kittens until they are sure of the best conditions, and a queen will spend considerable time looking for just the right place. She will then use it as a resting place for some days before her kittens are due, as if testing it for herself to make sure that it is what is needed, namely an enclosed space, on soft ground and dimly lighted.

Usually the domestic cat wants the birth-place to be secluded. Yet curiously, if there is a strong bond between the cat and her owner, the cat will do its best to ensure that the owner is in attendance at the birth. She will pester the owner, miaowing piteously until the owner accompanies her to the chosen place, delaying the moment of birth.

It seems that mares of the domestic horse can exercise some control over the timing of a birth and will do so if there is any kind of disturbance around them. This is, no doubt, a relic of the time when their ancestors were wild and beset by enemies.

For example, it has been found that the wildebeest mother can also delay the birth of her calf. The main enemies of the baby are hyenas which will try to carry off a newborn baby rather than one already able to run. Since a baby wildebeest can run seven minutes after birth this leaves the hyenas little time. The natural impulse of a wildebeest mother when hyenas attack is to make for the nearest herd. There her calf intermingles with other, older calves, making it difficult for the hyenas to single it out.

Innate behaviour, however, takes care of her actions. With hyenas around she delays the birth until she is near enough to the herd for the baby to stand a good chance of surviving its first vulnerable seven minutes and then reach the herd on its first tottering footsteps.

Pheromones, the external hormones

Pheromones are chemical messengers acting outside the body. Their action is extremely subtle and a typical example is the odour given off by the bitch when fully in heat and which can be perceived by a dog a hundred yards (90 m) or more away. Not only does the scent enter his nostrils, it also stimulates him to vigorous action.

The male red fox is the personification of selfish greed right up to the time when his vixen is about to have her cubs. She then retires to her specially prepared nursery and the dog-fox, instead of gobbling up all available food, as he usually does, takes it all to her. Moreover, he continues this unselfish behaviour until the cubs have been weaned. It can only be supposed that it is a pheromone emanating from the vixen that produces such a remarkable change in her mate. One further result of it is that should the vixen die before the cubs are independent the dog-fox will take over the task of fetching food and feeding them.

Tree shrews live in South-east Asia. They look like rats with bushy tails but behave like squirrels. They were once classified with monkeys and apes but they have now been placed in a category of their own. They are also distinct in their breeding habits. Being a fairly small animal and living in trees, the tree shrew builds a nest for its babies. It is the male, however, that builds the nest and he does so at the right time. To date it is not known whether this is due to the changed behaviour of the female as the moment of birth draws near. More likely is it that the female gives out an odour, a pheromone perhaps, that starts the male building a nest.

With the exception of the spiny mouse of Africa, all other baby mice and rats, and there are many different kinds throughout the world, are born in nests, naked, blind and helpless. The offspring of the spiny mouse are unique in this large family in being precocial (developed).

The spiny mouse has another peculiarity. Unlike most other mammals she does not seek solitude to give birth. She stays with her fellow spiny mice. Just before she is about to give birth she has an insatiable urge to lick something. She will, of course, turn this to good effect as soon as her babies are born, to lick them clean of their birth membranes and fluid. She will continue to lick them from then onwards to keep them clean.

Meanwhile, she has this urge to lick, so she tries to steal a baby from one of the other females, holding it in her paws. If unsuccessful she will seize the nearest adult and, holding it in her paws as she would a baby, lick it and groom it in a way she never would at ordinary times.

Something similar has been seen in a domestic cat about to kitten. Her companion in the household was one of her sons, now a well-grown tom-cat. Despite their blood-tie she could not bear him near her. She drove him away whenever he came near her as the time for birth of her next litter drew near. When she was about to have the litter, she sought him out, held him, and licked him all over. Presumably she was afflicted

with an excess of the appropriate hormone, whereas spiny mice seem to suffer from an excess of the hormone normally.

These last few examples seem somewhat unusual. Nevertheless, although there is a continuous thread running through the main story of the Family of Animals it does no harm to emphasize that innumerable diversions from it can be made. These do nothing to alter the fact that life as a whole can be seen as a series of cycles succeeding each other in an endless stream from birth to death, from generation to generation, each cycle bringing us back to the point at which this book started.

Egyptian spiny mouse mother carrying her two-day-old baby. Before her babies are born the female has a powerful maternal urge, and will even steal a baby from another mother.

Acknowledgments

The author's first acknowledgment must be of his indebtedness to his daughter Jane who, in addition to contributing a number of pictures, suggested the title and the theme of this book. She also undertook the onerous task of making the provisional selection of pictures, as well as contributing information whenever called upon to do so. My thanks are also due to Peggy Wratten for her extensive assistance in putting the text together, typing it, and helping to check the proofs.

Illustrations were supplied or are reproduced by kind permission of the following:

Afrique Photo: 18, 71 below (Norman Myers)
Ardea Photographics: endpapers (Clem Haager)
Australian News and Information Service: 49 below
S. C. Bisserot: 84–5
Black Star: 21 (Norman Myers)
Camera Press: 23 (Norman Myers)
Bruce Coleman: 53, 91, 112 left; 40–1 (H. Albrecht); 64 (Des Bartlett): 59 (S. C. Bisserot); 1, 2–3, 10–11, 13, 24, 26, 35 above, 45 above, 60, 66, 68, 69, 72, 86, 92, 93, 94 above and below, 95, 97, 101, 103, 106, 107, 109, 115 (Jane Burton); 17, 27, 70 (Bob Campbell); 44 (Norman Myers); 35 below (G. Pizzey); 29 (Fritz Polking); 47 (Masood Quarishy); 85 (D. Robinson); 112 right (R. Schroeder); 50–1 (J. Simon); 46, 83, 110 (Simon Trevor); 12, 96 (J. van Wormer)
Edistudio: 22 above; 38 (de la Fuente); 43 (Gutierrez)
F. Erize: 6, 80–1
H. Fristedt: 34
E. Hanumantha: 67
Hoa-Qui: 61
IGDA: 14, 15, 16 right
Jacana: 8, 71 above (A. Bertrand); 51 (A. R. Devez); 54 (A. Fatras); 37 (J. M. Fievet); 31 (P. Montoya); 52 (F. Roux); 32 (W. Schraml); 30 (Suinot); 45 below (J.-F. Terrasse); 91, 102 below (J.-P. Varin); 55 (F. Bel-G. Vienne); 81 (Wangi)
Aldo Margiocco: 16 left, 75
J. Markham: 28
NHPA: 62–3 (Andrew Anderson); 104 (K. B. Newman)
Okapia: 22 below, 56; 68–9 (A. Root)
Nancy Palmer Photo Agency Inc.: 99 (L. C. Bissell)
L. Pechuan: 30–1
Photo Researchers: 78 left (T. Angermayer); 98 (D. Hanley); 52–3, 82 (R. Kinne); 49 above (G. Leavens); 78 right (T. McHugh)
Popperfoto: 41, 46–7, 102 above (W. T. Miller)
S. Prato: 20
Prensa Española: 50 (Pato)
Masood Quarishy: 33
C. Rivero Blanco: 76
John Sacher: 4–5
Sirman Press: 36 (W. Scheithhauer)
Time-Life: 48, 65; 57 (R. Morse)
Tiofoto: 78–9 (S. Gillsätter)
Ag. Zardoya: 90
Zentrale Farrbild Agentur GmbH: 100

Index

Figures in italic refer to illustrations